AF756912

UNIPA Springer Series

Editor-in-chief

Carlo Amenta, Università di Palermo, Palermo, Italy

Series editors

Sebastiano Bavetta, Università di Palermo, Palermo, Italy
Calogero Caruso, Università di Palermo, Palermo, Italy
Gioacchino Lavanco, Università di Palermo, Palermo, Italy
Bruno Maresca, Università di Salerno, Fisciano, Italy
Andreas Öchsner, Griffith School of Engineering, Southport Queensland, Australia
Mariacristina Piva, Università Cattolica Sacro Cuore, Piacenza, Italy
Roberto Pozzi Mucelli, Policlinico G.B. Rossi, Verona, Italy
Antonio Restivo, Università di Palermo, Palermo, Italy
Norbert M. Seel, University of Freiburg, Germany, Germany
Gaspare Viviani, Università di Palermo, Palermo, Italy

More information about this series at http://www.springer.com/series/13175

Carmela Rita Balistreri

Endothelial Progenitor Cells

A New Real Hope?

UNIVERSITÀ DEGLI STUDI DI PALERMO

Springer

Carmela Rita Balistreri
Department of Pathobiology and Medical
Biotechnologies
University of Palermo
Palermo
Italy

ISSN 2366-7516 ISSN 2366-7524 (electronic)
UNIPA Springer Series
ISBN 978-3-319-55106-7 ISBN 978-3-319-55107-4 (eBook)
DOI 10.1007/978-3-319-55107-4

Library of Congress Control Number: 2017933060

© Springer International Publishing AG 2017
This work is subject to copyright. All rights are reserved by the Publisher, whether the whole or part of the material is concerned, specifically the rights of translation, reprinting, reuse of illustrations, recitation, broadcasting, reproduction on microfilms or in any other physical way, and transmission or information storage and retrieval, electronic adaptation, computer software, or by similar or dissimilar methodology now known or hereafter developed.
The use of general descriptive names, registered names, trademarks, service marks, etc. in this publication does not imply, even in the absence of a specific statement, that such names are exempt from the relevant protective laws and regulations and therefore free for general use.
The publisher, the authors and the editors are safe to assume that the advice and information in this book are believed to be true and accurate at the date of publication. Neither the publisher nor the authors or the editors give a warranty, express or implied, with respect to the material contained herein or for any errors or omissions that may have been made. The publisher remains neutral with regard to jurisdictional claims in published maps and institutional affiliations.

Printed on acid-free paper

This Springer imprint is published by Springer Nature
The registered company is Springer International Publishing AG
The registered company address is: Gewerbestrasse 11, 6330 Cham, Switzerland

This illustration summarizes the critical points relating to endothelial progenitor cells (EPCs), which are discussed in this monograph. In particular, the left hand side *shows a photo that emphasizes the fact that EPCs are considered by some researchers to be merely a dream.* The centre *of the figure shows a roadmap*—the clear puzzle of EPC data—*which is suggested and discussed in the Chap. 3. This roadmap requires a framework in order to shed light on the findings obtained to date on EPCs, since there is currently more confusion than understanding. This may permit fixing valid data, which may be useful as a cue for performing additional investigations. These steps may finally allow a real translation of EPCs to clinical applications,*

providing a real hope in the near future. In the figure, the hope is represented by the picture Hope of the Future of Eduardo Rodriguez Calzado *(*to the right*).* Eduardo Rodriguez Calzado *is an amazing artist and graphic designer born in 1976 in Torreon, Mexico. He studied Graphic Design in ISCYTAC, La Salle and furthered his studies at the Academy of Art College in San Francisco. Eduardo has a unique style of oil painting on canvas; he is an artist obsessed with details. His paintings capture the harmony of the universe filled with vibrant colours, geometric shapes, dashes and dots. Calzado creates striking abstract imagery in which he mostly represents the human form. He is inspired by tales heard from his family or friends; and by his own experiences and memories of childhood. As the artist says, his goal is to take the viewer to the world of harmony and fantasy.*

Dedicated to my family

Preface

Chronic inflammatory diseases, such as cardiovascular diseases (CVDs), diabetes, Alzheimer's disease (AD) and cancer, have a disproportionate prevalence with advancing age owing to the continuous growth in the aging population. This condition determines several medical, economic and social problems due to the dramatic increase in the number of affected individuals, who are not autonomous. Thus, research efforts are centred around reducing and/or delaying the onset and progression of these diseases by researching new strategies for early prevention and diagnosis. In this context, understanding the mechanisms involved in the tissue, organ protection and repair are imperative for the development of new preventive treatments. Accordingly, medical research is pursuing new ways of trying to face this imposing challenge, i.e. regenerative medicine with stem cells and progenitors, such as endothelial progenitor cells (EPCs). Since their discovery, EPCs have rapidly caught the attention of researchers for their ability to facilitate vascular repair in different ischemic tissues, by contributing to neovascularization in several tissue injury models. Interest has also heightened dramatically after evidence about their capacity to counteract related CVD endothelium dysfunction. In addition, recent studies, using different animal models of cancer, suggested the importance of bone marrow-derived EPCs (i.e. postnatal vasculogenesis) in tumor vascularization and growth. EPCs are present in the peripheral blood; their levels are increased in response to certain signals/cytokines; and they home into the neovascular bed of malignant tissues. Furthermore, at the clinical level, evidence is emerging that changes in EPC levels might predict the efficacy of anticancer drug combinations, such as anti-angiogenic agents. On the basis of these observations, EPCs have attractive potential diagnostic and therapeutic applications for malignant diseases. Additional recent evidence also suggests the possibility to adopt EPCs as prognostic biomarkers for AD. It has been observed that patients with AD have reduced circulating EPCs, suggesting that an anomalous capacity to regenerate endothelium is associated with AD. In the case of diabetes, numerous groups have detected decreased EPC numbers and functionality in affected patients, confirmed by decreased numbers of colony-forming units, decreased adhesion and migration and decreased tubule formation. In addition, it has been found that strategies based on the administration of statins, angiotensin converting enzyme inhibitors, angiotensin receptor blockers and peroxisome-proliferator-activating-receptor-γ agonists, up-regulate and enhance both the EPC number and functionality.

Emerging evidence also indicates that transplantation of EPCs is beneficial for the recovery of ischemic cerebral injury. EPC-based therapy could open a new avenue for ischemic diseases. Currently, clinical trials for evaluating EPC transfusion in treating ischemic stroke are underway.

However, much of the increasing evidence implicating progenitors in these diseases is contrasting. Thus, their real role remains uncertain. This is compounded by the necessity for a standardization of the different methodologies and protocols for characterizing, identifying and defining these cells, or their subsets. This problem represents one of the major consequences of the large heterogeneity that exists in data from the literature. In this monograph, some of these aspects are discussed using research to give clear indications regarding EPC functions and definitions, as well as evidence to support the problem of their characterization. In addition, recent findings on their role as disease biomarkers and exogenous or autologous cell therapy are provided. Certainly, limitations are also stressed.

Based on these observations, this monograph, structured in 3 chapters, will describe, in the first chapter, the relevance of translational medicine (TM) as a new research approach to counteract the imposing challenge of age-related diseases. In particular, a considerable emphasis will be given to regenerative medicine (RegMed), a new branch of TM. RegMed can be used to improve health and quality of life, by restoring, maintaining or enhancing tissue and functions of organs. Of the diverse RegMed approaches, a particular focus will be given to stem/progenitor cell-based therapies, their benefits and disadvantages, as well as to the description of types of stem and progenitor cells considered for regenerative cell therapies, such as EPCs as emerging candidates for RegMed applications. To understand the value and potentiality of EPCs in this field, as well as the related limitations, in Chap. 1, the biological features, origins, sources and endogenous functions of EPCs will be reported. In addition, a critique of EPCs, namely relating to their characterization and definition, will be put forward, because they represent the real cause of the extensive heterogeneity of existing data in the literature on this topic, which creates more confusion than comprehension. Chapter 2 outlines the clinical relevance of EPCs as both potential predictors, and diagnostic and prognostic biomarkers of age-related diseases and therapeutic agents, by discussing the advantages, disadvantages and the conflicting data. To conclude, Chap. 3 will propose a potential roadmap for revising the findings obtained to date, reordering them and creating a clear puzzle of the valid data. It can facilitate understanding of important aspects, including firstly defining, isolating and characterizing EPCs by establishing a standardized criteria for EPC research, discrimination of appropriate sub-populations for cell therapy, timing, dosing, priming of cells, and delivery mode for different applications. Furthermore, influencing factors might also be identified with the aim to focus resources and efforts, and the problems linked to the local retention and fate of cells in the therapeutic target zone might be resolved. Finally, an overview of innovative strategies will be reported, which might improve the efficacy of cell therapy at all levels, including cell priming, bio-nanotechnology, and tissue engineering. They might be used as emerging tools, and, in particular, their combination might produce very advantageous results.

Such descriptions and discussions address an extensive audience, including students, health care professionals, biologists, physicians and diverse scientific community. In particular, it will be a valuable resource for clinical scientists, researchers, university professors, health practitioners, gerontologists and geriatricians, students, and for all those who wish to broaden their knowledge in the allied field. Policy makers and agencies involved in implementing preventive policies might also use this monograph as an updated integral resource. All government and private organizations, including libraries at the college level, academic universities, and research institutions might benefit from it as resource for its current references.

Palermo, Italy
July 2016

Carmela Rita Balistreri

Acknowledgements

A heartfelt thanks to Prof Calogero Caruso: Renowned Researcher, Mentor and Friend. This work has been supported by grants from the Italian Ministry of Education, University and Research to Dr. Balistreri CR and Prof. G. Mazzesi (000134_RDB_2014 "Focus on cellular and molecular mechanisms of both, protection and repair in BAV vs. TAV patients as novel tools for personalized treatments").

Contents

About the Author

Dr. Carmela Rita Balistreri received her Ph.D. in Pathobiology in 2007. She is an Assistant Professor of Clinical Pathology, and a member of the Department of Pathobiology and Medical Biotechnologies, University of Palermo, Italy. In the first phase of her studies, Dr. Balistreri was actively involved in the study of the immunological mechanisms of senescence, such as apoptosis, interleukin production, and the activity of natural killer and granulocyte cells in aged individuals. After this, her main field of scientific interest was the study of the association between the hemochromatosis gene mutations and longevity, and age-related diseases, such as acute myocardial infarction and sporadic Alzheimer's disease. Subsequently, her principal interest has been focused on biogerontological studies, including case control studies with the key aim of analyzing the role of candidate immune/inflammatory genes (i.e, TLR4, TLR2, CCR5, CRP, Connexin37, MMP9, PECAM-1/CD31, CD14, Cox and 5LO) in longevity and age-related diseases (i.e., cardiovascular diseases, Alzheimer's disease and prostate cancer). In these studies, a particular approach has been used: centenarians have been used as a second control group, the "supercontrol" group, since they have escaped major age-related diseases. In particular, these studies demonstrate the role of anti-inflammatory polymorphisms of genes, involved in inflammatory and innate immunity responses, to increase the chance of achieving longevity in a modern environment with reduced pathogen load and improved control of severe infections by antibiotics. Furthermore, these data also evidenced the opposite role of these polymorphisms in longevity and age-related diseases, including myocardial infarction, prostate cancer and Alzheimer's disease. In the specific field of Alzheimer's disease, she has published several papers with her group on the role of genetics on inflammation in AD pathophysiology. Recently, Dr. Balistreri has also focused her studies on identifying the cellular, molecular and genetic mechanisms involved in the pathophysiology of thoracic aorta aneurysms and mitral valve diseases, suggesting diagnostic, prognostic and preventive biomarkers and targets for personalized treatments. Dr. Balistreri is also performing studies on endothelial progenitor cells as potential disease biomarkers of some cardiovascular diseases, such as bicuspid aortic valve. International recognition of her scientific achievements is reflected in frequent citations of her papers (http://www.scopus.com) and her referee activity for the several journals, including *Scientific Reports, Human Immunology, Diabetic Medicine, Expert Review of Neurotherapeutics, Human Genetics, Ageing Research Reviews, Circulation, Immunity and Ageing,* and *PLOSone*. Her h-index for the 96 papers (1998–2016) is 25, (http://www.scopus.com; orcid.org/0000-0002-5393-1007). To improve her scientific cultural level and to compare her data with those of other groups, she has attended 80 congresses focused on following topics: Age-related diseases, Immunosenescence, Longevity, Model system, Aging & Wellness and Regenerative Medicine, Cardiovascular diseases. She has attended several congresses as a speader. Dr. Balistreri has received several national and international awards and honors.

Acronyms

7-AAD	7-aminoactinomycin D
ACS	Acute coronary syndrome
AD	Alzheimer's disease
Ang	Angiopoietin
APP	Amyloid precursor protein
AR	Aortic valve regurgitation
ASCs	Adipose stem cells
AT-MSCs	Adipose tissue-derived mesenchymal stem cells
Aβ	Amyloid beta protein
BDNF	Brain derived neurotrophic factor
BM	Bone marrow
BMSCs	Bone marrow stem cells
CACs	Circulating angiogenic cells
CAD	Coronary artery disease
CECs	Circulating endothelial cells
CEPs	Circulating endothelial progenitors
C-kit	Proto-oncogene c-kit
CPCs	Circulating H-progenitor cells
CREB	cAMP response element binding
CSCs	Cardiac-derived stem cells
CVDs	Cardiovascular diseases
CXCR4	Leukocyte-derived seven transmembrane-domain receptor-4
ECs	Endothelial cells
EDTA	Ethylenediaminetetraacetic acid
eNOs	Endothelial NO synthase
EO-CFUs	Early-outgrowth colony-forming units
EPC	Endothelial progenitor cell
EPO	Erythropoietin
ESA	Erythropoietin stimulating agents
ESCs	Embryonic stem cells

FGF2	Fibroblast growth factor 2
FMO	Fluorescence-minus-one
FSC	Forward scatter
GCSF	Granulocyte-colony stimulating factor
GIST	Gastrointestinal stromal tumor
GSK-3β	Glycogen synthase kinase-3β
H_2O_2	Hydrogen peroxide
HCC	Hepatocellular carcinoma
HPCs	Hematopoietic progenitor cells
HSCs	Hematopoietic stem cells
IL-10	Interleukin-10
IL-1β	Interleukin-1β
iPSCs	Induced pluripotent stem cells
ISHAGE	Hematotherapy and Graft Engineering
MI	Myocardial infarction
MMP-9	Metalloproteinase-9
MSCs	Mesenchymal stem cells
NGF	Nerve growth factor
NH_4Cl	Ammonium chloride
NO	Nitric oxide
PBMCs	Peripheral blood mononuclear cells
PSGL-1	Glycoprotein ligand-1
RCC	Renal cell carcinoma
RegMed	Regenerative medicine
ROS	Reactive oxygen species
SCD	Stable coronary disease
SDF-1	Stromal cell derived factor 1
SPC	Stem progenitor cell
T2DM	Type 2 diabetes mellitus
TM	Translational medicine
TNF-α	Tumor necrosis factor-α
VEGF	Vascular endothelial growth factor
VEGFR2	Vascular endothelial growth factor receptor 2

Chapter 1
From Regenerative Medicine to Endothelial Progenitor Cells as Potential Candidates

Abstract A promising branch of translational medicine is *Regenerative Medicine* (RegMed). RegMed aims to improve health and quality of life by restoring, maintaining or enhancing tissues and functions of organs. In order to achieve this, RegMed utilizes several approaches. Among these, *cell-based therapy* represents the central focus of RegMed. Several cellular types have been considered for the investigation and application of regenerative cell therapy in disease organs. and bone marrow (BM)-derived endothelial progenitor cells (BM EPCs) are emerging as potential candidates. Here, particular emphasis is given to the description of their biological features, for understanding their relevance and potential in this field, as well as the related limitations in this and in successive chapters.

Keywords Regenerative medicine · Cell-based therapy · Stem/progenitor cells · Endothelial progenitor cells

The above figure illustrates *the myth of the Titan Prometheus and is an icon for Regenerative Medicine. Prometheus was known as the benefactor of mankind for his desire to assist mortals and give them many beneficial gifts that helped them to survive and to live prosperous lives. His duty as a god was to form man from water*

© Springer International Publishing AG 2017
C.R. Balistreri, *Endothelial Progenitor Cells*, UNIPA Springer Series,
DOI 10.1007/978-3-319-55107-4_1

and earth. In doing so, he gave them each a gift of strength or speed, craftiness or wisdom, and many other gifts that improved their ways of living. The most well known gift from Prometheus was that of fire, which Prometheus stole from Zeus' lightning bolts. Prometheus' defiance and betrayal provoked Zeus to have him chained to Mount Caucasus, where an eagle ate daily from his ever-regenerating liver. As an immortal Titan, Prometheus' liver grew in the night what the eagle had eaten the day before. His torture lasted 30,000 years until he was freed by Hercules.

The myth of the great Titan Prometheus, as introduced by Hesiod in his Theogony (eighth century B.C.) provides an icon for Regenerative Medicine. When he stole fire, a symbol for civilization and technology, to give it to mankind, he received a cruel punishment. Jupiter had him chained to the Carpathian Mountains, where an eagle by the name of Ethon would pick at his liver every day; his liver would then regenerate during the night (see Figure). His torture lasted 30,000 years until he was freed by Hercules. The ancient Greeks were aware of the regenerating capacity of the liver, hence they named it 'hepar' after hepaomai, meaning to 'repair oneself'

The brief description of this historical event and the icon of regenerative medicine (RegMed) allows the introduction of a critical analysis of the actual status of the field of RegMed, and it co-temporally constitutes the filter through which to put together a clearer picture of its future. As it is well recognized, until now RegMed does not determine the therapeutic revolution, which was awaited already before its birth. After about two decades of extremely high expectations and often disappointing returns, both in the medical as well as in the financial arena, this scientific field reflects the sense of a new era and suggests the feeling of making a fresh start, although many scientists are probably seeking reorientation. Much of research was industry driven, so that, especially in the aftermath of the recent financial meltdown, it witnessed a biotech asset yard sale. Despite any monetary shortcomings, from a technological point of view, there have been great leaps forward that are yet to find their way to the patient. RegMed is definitely bound to play a major role in our lives because it embodies some of the primordial dreams of mankind: everlasting youth, flying, remote communication and setting foot on the moon. Scientific journals have been the voice of these developments in RegMed from its beginning and actually reflect the recent scientific advances in this field. Therefore, the idea of using this figure, it stays in trying to look at RegMed through the keyhole of history, which might just be like looking '*back to the future*'. Thus, the principal message of this monograph suggests that "we are almost there", in terms of being able to produce tissue replacement "off the shelf", in the near future for everyone in need. However, it seems— literally—that we are able to fabricate constructs that "look like tissue, smell like tissue and taste like tissue" but not some that also function equally like tissue. Thus, a real therapeutic evolution into the health and clinical sectors still remains to be observed and applied. More critically, others have proclaimed that the time must come to "stop tissue engineering and start engineering tissues". In order to achieve this goal, the following measures are needed: (1) to consider the advances and the limitations of this field; (2) to re-examine the data obtained and the associated controversies (3) to filter them and (4) to put together only the valid parts (5) for constituting a clear puzzle by using standardized research methods and methodologies; (6) this will be useful for pursuing innovative ways through a strong cooperation among scientists, physicians, administrators and public officials. Likewise, the same revision's approach might be used for literature data about bone marrow derived endothelial progenitor cells, as stressed in this monograph. The obstacles may be many, and the road may be hard and long, but the advances obtained may offer us greater and unique opportunities to meet the considerable challenge.

—*Carmela Rita Balistreri*-

1.1 Regenerative Medicine: A New Branch of Translational Medicine

Multi-factorial age-related diseases (i.e. cardiovascular diseases (CVDs), Alzheimer's disease (AD), diabetes and cancer) have a disproportionate prevalence in Western populations. This is a result of a variety of causes, including the continuous growth of aged population. Barring major unpalatable political changes, this current trend will undoubtedly continue. It is estimated that, by 2030, the old population will comprise between 36.9 and 40.5% of the entire population, among both Europeans and Americans. As result, there will be an even greater demand for pharmacological and invasive therapies in the future. In addition, the efforts performed to date by preventive clinicians and epidemiologists in researching real solutions to multi-factorial age-related diseases (e.g., preventive measures) have been limited in terms of their realization and application because they are insufficiently valued by the public, the media, and by health policy authorities or legislators. As result, refreshing and alternative directions in both preventive measures and biomedical research are of the upmost importance.

In recent years, biomedical research has pursued new ways of addressing this considerable challenge. In particular, biomedical research has evolved into an innovative data-intensive science that involves the collection of prodigious amounts of data from diverse resources at any given time. The value and relevance of these data are then only acquired through appropriate analyses, which ultimately results in knowledge that can be translated to various applications. Such an approach characterizes *translational medicine,* an emerging area comprising multidisciplinary research and scientific discoveries. Consequently, translational medicine is considered an innovative tool for improving human health, through reducing disease incidence, morbidity and mortality. Indeed, TM aims to improve health by translating findings from basic research studies into practical applications in terms of ways to prevent, diagnose or treat disease (i.e. *personalized therapies*) (Abraham et al. 2012; Wang 2012).

From the various domains of translational medicine, a new branch, RegMed, has emerged as being a necessity for closing the gap between the increased risk of disease and the decreased capacity of the majority of human tissues and organs to spontaneously regenerate and to respond to insults and damage among aged individuals. RegMed is an interdisciplinary field of research and clinical applications focused on the repair, replacement or regeneration of cells, tissues or organs to restore impaired function resulting from any causes, including congenital defects, diseases, trauma and aging. It is, indeed, proposed as the way "*to improve the health and quality of life by restoring, maintaining or enhancing tissue and functions of organs*." In order to achieve these objectives, RegMed uses a combination of several technological approaches, which move it beyond traditional transplantation and replacement therapies. They include (but are not limited to) the use of soluble molecules, gene therapy, stem cell transplantation, tissue engineering and the reprogramming of cell and tissue types (Mason and Dunnill 2008a, b; Atala 2012). In particular, significant advances in the fields of stem cell biology and bioengineering have converged within RegMed (Fig. 1.1). They have led RegMed

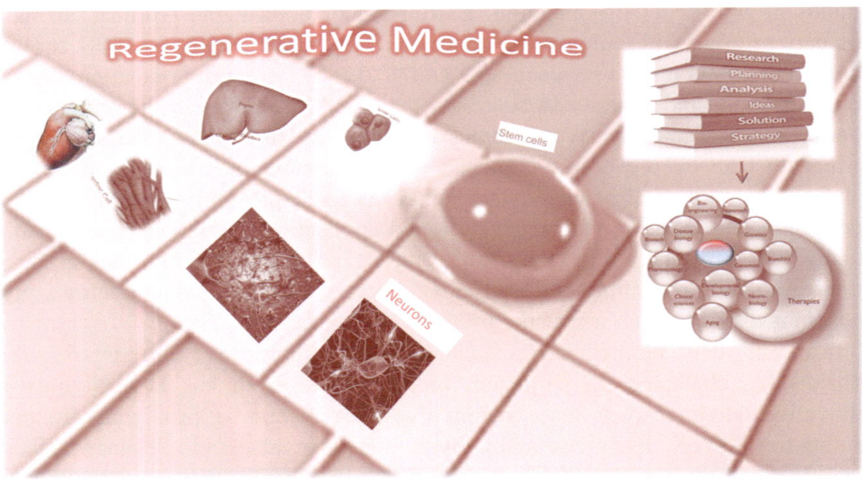

Fig. 1.1 Regenerative medicine. An emerging branch of translational medicine focused on the repair, replacement or regeneration of cells, tissues or organs to restore impaired function resulting from any causes, including congenital defects, diseases, trauma and aging. Stem cells are the cornerstone at the heart of regenerative medicine and might provide the potential solution for human diseases

scientists to develop *cell therapies* based on the capacity of stem cells to restore damaged tissues, by transplanting tissue patches, with a preserved 3D structure. However, RegMed scientists also utilize other natural human substances, such as genes, proteins, vesicles, biomaterials and nanotechnologies to regenerate diseased or damaged human tissues and to restore normal functions. This is in agreement with RegMed goal, based on the concept that it is not just to replace what is malfunctioning, but to provide the elements required for *in vivo* repair; to devise replacements that inconspicuously interact with the living body and stimulate the body's intrinsic capacities to regenerate (Atala 2012; Rouchi and Mahdavi-Mazdeh 2015).

The following section describes stem cell-based therapy, which remains the central focus of RegMed, and stresses its benefits and disadvantages.

1.1.1 Stem Cell Therapy

As previously mentioned, stem cell therapy represents a new class of medicine, and it is one of the approaches of RegMed. Ever since their discovery and initial isolation, the scientific community has embraced stem cells for their potential to change the therapeutic approach to chronic diseases, by moving from disease management toward a regenerative framework. However, it is necessary to clarify that this strategy is not new, since it has already been successfully utilized in the clinic in the form of organ transplantation. Organ transplantation provides the possibility to replace malfunctioning organs with unscathed donor organs, which are able to take over the functions and to improve and extend the life's quality of the patients. However, this approach is severely hindered by the shortage of donor organs, and, therefore, it will never become a viable medical treatment option for the majority of patients. Thus, progression and refinement in terms of both the isolation and culturing procedures of stem cells over recent decades have reopened the door for RegMed approaches, in hope of healing the chronically damaged organs of patients who are on a transplant waiting list (Rouchi and Mahdavi-Mazdeh 2015).

In this regard, it is imperative to underline the fact that cell-based therapy has features that distinguish it from drug therapy, or other types of RegMed therapies. Cell-based therapy is based on cells, and they represent the most complex biopharmaceuticals. Protein or gene therapies are based on relatively simple macromolecules, and they are ideally suited to target a single defect, rather than eliciting a complex biological regenerative response (for which stem cells seem to be better suited). However, cells are more complex to utilize. Firstly, they are composed of thousands of proteins and an entire genome. In addition, they are dynamic and able to react to their environment. Upon transplantation, the transcriptome, proteome, and even secretome profile of cells can change, thereby altering their function compared with that observed *in vitro* upon initial culture expansions. All this suggests, on one hand, that cells are a unique multidimensional therapeutic treatment well-suited to the RegMed approach. On the other hand, it also suggests that cells are a complex and challenging entity to utilize and study (Rouchi and Mahdavi-Mazdeh 2015).

As above underlined, the cell therapies have unique proprieties when compared with other RegMed approaches or standardized drug treatments. This renders difficult to find an appropriate pharmacokinetics for each type of cell therapy, which is fundamental for its final result. Firstly, the methods of *administration* differ to those used for a drug. The preferred drug route is oral administration. For cell therapies, this route is not advantageous, because cells are not capable of surviving in the acid environment of the stomach, and even less so in the intestine. A more suitable administration route for cell therapies might be into the circulation, by means of an intravenous injection, as for the hematopoietic stem cells (HSCs) therapy in cases of leukemia treatment (Thomas and Epstein 1965). This way could also be used for cardiac cell therapy, considering the hypothesis that it could rely on similar chemo-attractant signals coming from the damaged heart to attract the circulating cells to the injury site. Although this administration route is the most deliverable and cost effective (Boyle et al. 2006), the cells are faced with many obstacles prior to their arrival in the specific damaged organ, such as the heart. In this case, a more local delivery strategy, such as the intracoronary injection, allows the administration of stem cells directly into the cardiac coronary circulation (Chen et al. 2004). In such an approach, the injected cells are faced with high coronary flow and are required to transmigrate through the capillaries for repairing the myocardium. Thus, an alternative approach is the administration route into the myocardial tissue, despite the fact that cell spillage can occur at the site of injection (Sherman et al. 2006). In addition, it can be difficult to bring the cells to a preferred location in order to avoid injecting them into a remote region that is too far from the injury, or into an ischemic area deprived of oxygen and nutrients (Feyen et al. 2016).

The pharmacokinetics of cell-based therapies is characterized by other difficulties including: (1) *the inability to monitor the bio-distribution of stem cells after their administration*; (2) *the survival of stem cells in the damaged tissues, which is very difficult* (cellular therapeutics delivered into ischemic myocardial tissue arrive in a hostile inflammatory milieu, and are therefore susceptible to pro-apoptotic signaling); and (3) *engraftment and integration* (in case of cardiac stem therapy, engraftment is only the first step toward remuscularization, since subsequent organization and proper tissue integration are critical for the participation of engrafted cardiomyocytes in heart repair) (Feyen et al. 2016).

These limitations have led the researchers to develop delivery strategies to enhance the retention, survival and integration of stem cells. Approaches utilizing pharmacology, genetic manipulation, biological or material incorporation have been implemented to improve these aspects. However, the pharmacokinetic proprieties of cells are not the unique considerations for the further improvement of cell-based therapies, since therapeutic action can also be addressed by modulating the performance of stem cells.

All these observations underline the fact that the early promise of cell therapy to repair the damaged organ by injecting different cell types has not yet been fulfilled.

Whereas cell therapy was initially suggested as a method to help and repair the damaged organ, current thoughts on cell therapies are mainly focused on the concept that the injected cells produce and secrete a pleiotropic repertoire of factors. These factors influence the cellular environment and thereby help the stressed tissue. As result, cellular therapy becomes a complex delivery tool for reparative biological drugs. First-generation cells, including BM stem cells and their secretomes, are mainly aimed at cellular salvage and at stimulating the endogenous repair mechanisms of the damaged organs through pro-angiogenic or pro-survival activities. However, the utilization of these cells in the clinic has never reached the awaited level of results. Future studies will need to address the underlying poor pharmacokinetic properties, in order to bolster the effects of these first-generation paracrine therapies. Furthermore, the manipulation of cells to rejuvenate the patients' own cells or to enhance paracrine action are also being explored. This will lead to the application of cell therapy as a complex delivery tool in which the slow release of reparative signals is further enhanced. The paracrine effects will boost endogenous repair mechanisms to maintain organ homeostasis.

Overall, a shift from the initial pragmatic delivery approaches toward tailored delivery strategies aimed at improving pharmacokinetics by using pharmacodynamics properties, which will consent their clinical application. However, once optimal strategies are developed, they can be coupled with technologies that have been developed over the last decade of stem cell research, including cardiac cell therapy, injection catheters and imaging modalities. Furthermore, recent clinical trials have helped to train clinicians for the delivery of cells into the myocardium and lay a foundation for cell therapy work in many medical centers around the world. These advances will help expedite the transition of future cell therapies toward patients. Although cellular therapeutics have failed to survive to their initial hype, careful re-evaluation of their mode of action and steps to address the current pitfalls should help to unlock the vast potential of stem cells, and help them to reach the clinic in a timely fashion.

Furthermore, the research community is also focusing its attention on progenitor cells as optimal candidates. Among these, BM-derived endothelial progenitor cells (EPCs) are emerging as candidates for several applications. In this monograph, the potential relevance of cell-therapy based on use of BM-EPCs will be discussed. In addition, particular emphasis will be given to its application in the cardiovascular field, using it as template for describing its use in other medical areas to treat other disease organs. Finally, the inconsistent evidence in this arena, which has hampered its effective clinical application in the management of several age-related diseases, will also be stressed.

In the following section, some relevant concepts will be highlighted, before discussing EPCs, in order to improve understanding of their features.

1.1.2 *Stem Cells as Appropriate Candidates of Cell Therapy: Types of Stem Cell*

To carry out cell-based therapy, stem cells and progenitors have been considered. For a better understanding this selection, it is well to point what it intends for stem cells. Stem cells are defined as undifferentiated cells with the potential to renew themselves, and to differentiate into any other specialized cell of human body, and, therefore (potentially and theoretically), to create any tissues or organs. Under specific conditions, stem cells can, indeed, differentiate into a diverse population of mature and functionally specialized cellular types. To date, different classes of stem cells are recognized. Among these, in the first class, there are the totipotent cells, which have the capacity to differentiate into embryonic and extra embryonic cell types, thereby generating entire organisms, even if this capacity is limited to cells produced by the first few divisions after fertilization. Pluripotent stem cell types are another class. They can differentiate into all embryonic cell types and form ectoderm, endoderm and mesoderm cell lineages, but not into extra embryonic cell types, and thereby they can form all the cell types of an adult organism (Fig. 1.2). Finally, there is the class of adult multipotent/unipotent stem cells, often termed progenitor cells, which can only differentiate into several closely related cell types (Stoltz et al. 2015) (see some examples in Fig. 1.3).

A variety of cellular types have been and are currently used in RegMed, including BM-derived mononuclear cells (BM-MNCs), peripheral blood mononuclear cells (PBMCs), mesenchymal stromal cells (MSCs), embryonic stem cells (ESC), induced pluripotent cells (iPSCs), and organ-specific stem cells [i.e. such as

TYPES OF STEM CELLS

Type of stem cell	What it can be	Examples
Totipotent cells	Each cell can develop into a new individual	Cells of embryo of 1-3 days
Pluripotent cells	Each cell can form any cell type (over 200)	Cells of blastocyst 5-14 days
Multipotent cells	Cells differentiate and can form a number of tissue types.	Fetal tissue, cord blood, adult cells

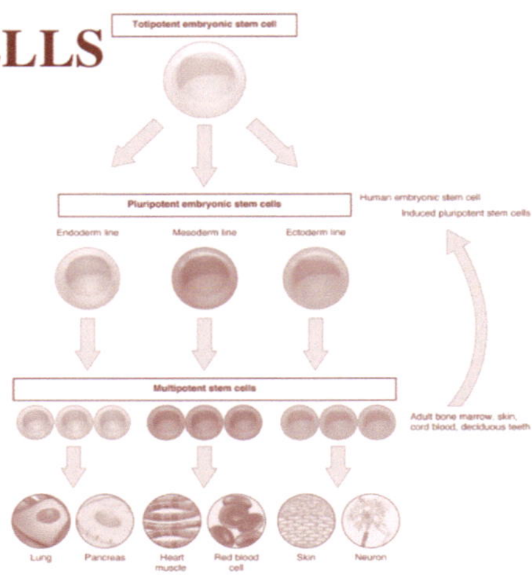

Fig. 1.2 Types of stem cells. As reported, there are several types of stem cells with different features

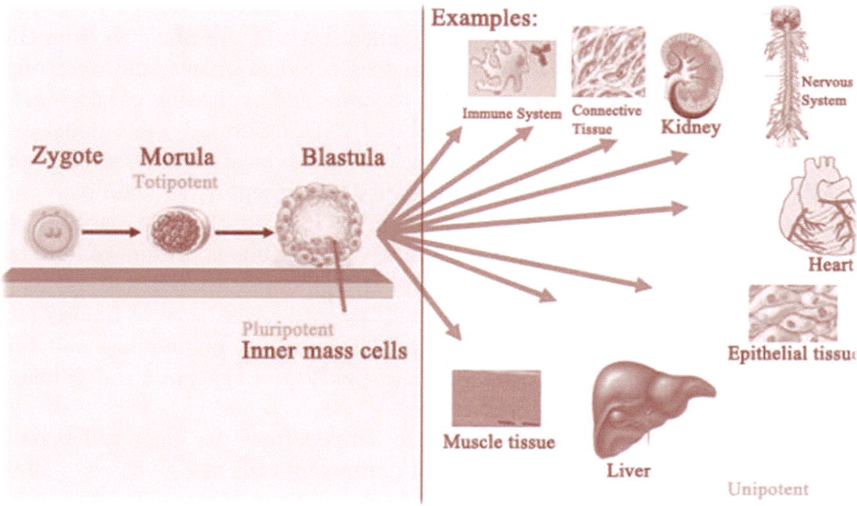

Fig. 1.3 Unipotent stem cells. In organs (e.g., heart, liver and kidney) and some tissues (e.g., epithelial and connective tissues), there are unipotent stem cells involved in their homeostasis and repair

liver- and cardiac-derived stem cells (CSCs)]. A brief description of these cells follows:

- *Bone marrow-derived mononuclear cells (BM-MNCs)*: BM is present in the interior of bones, mainly facilitating hematopoiesis and the production of several immune cells. In this mixture of cell types, several cell populations are present that have progenitor cell characteristics and properties. These cells are usually obtained by BM aspiration and purification via special techniques such as Ficoll–Paque-based separations, to obtain the mononuclear fractions of BM. This mixture of cells has several pro-reparative capabilities and is usually used directly upon purification.
- *Peripheral blood mononuclear cells (PBMCs)*: The population of mononuclear cells derived from blood contains a wide range of progenitor cells that have been used in cardiac cell therapy. Often, PBMCs are harvested after mobilization with granulocyte colony stimulating factor (G-CSF). Upon injection into the body, the G-CSF glycoprotein stimulates progenitor cell release from the BM into the blood stream where they can thereafter be isolated in a less invasive manner compared with BM aspirations (Pompilio et al. 2004; Boyle et al. 2006). After Ficoll–Paque separations, the stem/progenitor cell populations can be further enriched by CD34+ or CD133+ selection (Pompilio et al. 2004; Boyle et al. 2006).
- *Mesenchymal stromal/stem cells (MSCs)*: MSCs are multipotent stromal cells with a predominantly mesodermal origin, and are one of the most translational

attractive progenitor cell types, since these cells can be relatively easily isolated and expanded *in vitro* from individual patients. MSCs are obtained from BM, peripheral blood, adipose tissue or other mesenchymal organs, and were originally defined by their plastic adherent properties and expression of specific cell surface makers, such as CD105, CD90 and CD73. Currently, the global definition of MSCs is overly simplistic, since differences have been observed among MSC populations derived from different tissues, presenting an additional challenge to devising a universal definition. MSCs possess the capacity for self-renewal and the phenotypic potential to adopt a spectrum of different somatic cell types, including the osteoblast, adipocyte, endothelial and chondrocyte lineages (Pittenger and Martin 2004). Moreover, several reports have demonstrated that these cells have modulatory effects on immune cells and thereby are considered to have immune privilege (Aggarwal and Pittenger 2005).
- *Embryonic stem cells* (*ESCs*): ESCs are derived from the inner cell mass of blastocysts (an early embryo). ESCs are pluripotent cells and give rise to all the derivatives of the primary germ layers, the ectoderm, endoderm and mesoderm, and possess a high level of clonality, self-renewal and pluripotency. ESCs can develop into each of the more than cell types of the adult body when stimulated (with the correct cues). These cells express several transcription factors, including Oct-4, Nanog, and Sox2, which ensure the suppression of genes that lead to the differentiation and maintenance of pluripotency (Aggarwal and Pittenger 2005). Undifferentiated ESCs have been shown to induce teratomas and immunological rejection after injection into the myocardium, for example (Nussbaum et al. 2007).
- *Induced pluripotent stem* (*iPSCs*): Induced pluripotent stem cells are not adult stem cells, but rather reprogrammed adult cells (e.g., epithelial cells) that give rise to pluripotent cells. By using genetic reprogramming with transcription factors Oct-3, -4, Sox2, c-Myc and Klf4, pluripotent stem cells, which are equivalent to ESCs, have been derived from human adult skin tissue. Like ESCs, iPSCs can also be differentiated into all three germ layers. These cells can be generated from human and mouse somatic cells and thereby overcome ethical and immunological issues such as those identified for ESCs since both individual informed consent and patient-specific cells can be obtained. Despite the fact that iPSCs have an advantage over ESCs by avoiding ethical and immunological issues, significant safety concerns currently limit their clinical applicability (Okano et al. 2013).

Among these cells, ESCs and iPSCs exhibit nearly unlimited potential to differentiate *in vitro* and *in vivo*, but their applications are limited by ethical, legal and political concerns, as well as by scientific and clinical issues of safety and efficacy (Stoltz et al. 2015). Therefore, tissue-specific stem cells derived from adults offer alternative approaches that circumvent many of these concerns.

However, stem cells for RegMed applications should be consistent with the following criteria (Stoltz et al. 2015):

- Can be found in abundant numbers
- Can be harvested by a minimally invasive procedure with minimal morbidity
- Can be differentiated along multiple cell lineage pathways in a controllable and reproducible manner
- Can be safely and effectively transplanted to either an autologous or allogeneic host
- Can be produced in accordance with current "Good Manufacturing Practice" guidelines.

Among adult stem cells (Fig. 1.4), haematopoietic stem cells (HSCs), and, in particular, BM-MNCs, have been the focus of intensive research, and have been proven to be invaluable and life saving following inadvertent radiation exposure and in the treatment of hematologic malignancies. In addition to BM, HSCs can also be isolated from umbilical cord blood and from cytokine mobilized peripheral blood, making them readily available in relatively large quantities, and making relatively non-invasive collection possible (Porada et al. 2015). The ability of these cells to completely repopulate the entire hematopoietic system would already guarantee them a valuable place in the field of RegMed. However, the finding that hematopoietic chimerism can induce immune tolerance to solid organs and correct autoimmune diseases, such as systemic lupus erythematosus (SLE), multiple sclerosis and diabetes, has further broadened the clinical utility of HSCs (Porada et al. 2015). Their application has increased further in recent years, with the demonstration that they, through a variety of mechanisms, promote repair/regeneration of several tissues, such as the liver, heart, brain and others. However, their use in clinical applications is limited by harvest procedure. The harvesting of HSCs by BM aspiration is a painful procedure and the number of cells acquired is usually low. In addition, it may determine morbidity. These limitations have been and are leading to preclinical studies being performed in small and large animals, in order to carry out human clinical trials using the same cell therapies in the near future. However, contrasting results have been (and are being) obtained from diverse animal models. A significant suggestion, which has emerged as a result of comparison, is that the data from mice are likely more reliable. Thus, mice actually remain the absolutely essential and invaluable tools for both understanding basic HSC biology and developing novel therapeutic approaches to treat human diseases. Nevertheless, they are not simply "little humans", and the promising results obtained in murine models do not always translate directly into therapeutic successes in the clinical arena. As such, it is essential to identify and make use of an animal model system that most closely approximates the specific human physiological and/or disease processes, when performing preclinical safety and efficacy studies (Porada et al. 2015).

As an alternative and more efficient candidate for RegMed applications, human adipose tissue has been suggested to be an optimal source of stem cells, since it can be collected in large quantities (Lindroos et al. 2011) (Fig. 1.5). Human adipose stem cells (ASCs) are an abundant cell source with therapeutic applicability in pre-clinical studies in diverse fields, due to their ability to be readily expanded and

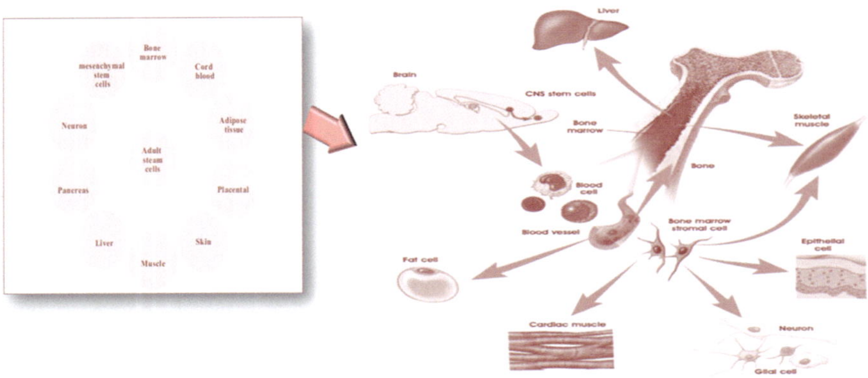

Fig. 1.4 Adult stem cells and their localizations

their large capacity to differentiate *in vitro* in several cell types, from the adipogenic type to osteogenic, chondrogenic and neurogenic varieties. Furthermore, ASCs have been shown to have immune-privilege and to be more genetically stable in long-term culture, when compared with BMSCs. The safety and efficacy of ASCs for tissue regeneration or reconstruction is currently under assessment in clinical trials (Lindroos et al. 2011). The number of trials investigating the efficacy of treating conditions such as Type I and II diabetes, liver cirrhosis and regeneration, fistulas, CVDs, limb ischemia, amyotrophic lateral sclerosis and lipo-dystrophy have risen rapidly, even if a very limited number has been completed (http://clinicaltrials.gov). Furthermore, ASCs are also under examination in clinical case studies for graft-versus-host disease, immunosuppression (rheumatoid arthritis, Crohn's disease and ulcerous colitis), multiple sclerosis, soft tissue augmentation and bone tissue repair. Clinical bone tissue reconstruction studies using autologous ASCs are also ongoing (Lindroos et al. 2011).

Currently, other potential candidates for RegMed cell therapy are emerging, including progenitor cells from BM or other tissue niches. Among these, EPCs are actually emerging as new therapeutic agents for several age-related diseases. They also represent the most widely studied adult human progenitor cell subpopulation up to now. The interest of the research community on EPCs arises from advances, over the last decade or so, relating to the discovery of postnatal vasculogenesis (called neo-angiogenesis), which is brought about by circulating progenitor cells, capable of differentiating into mature blood vessel endothelial cells. Thus, EPCs and their biology have became a common point of interest for physicians and basic scientists with the ultimate goal of translating their research into clinical application by using the innate reparatory mechanisms of the heart and vascular endothelium, as well as of other organs.

In this monograph, EPCs represent, as previously mentioned, the fundamental object of both description and discussion, and their potential as emerging therapeutic agents (as above mentioned) for CVD and other age-related diseases and for

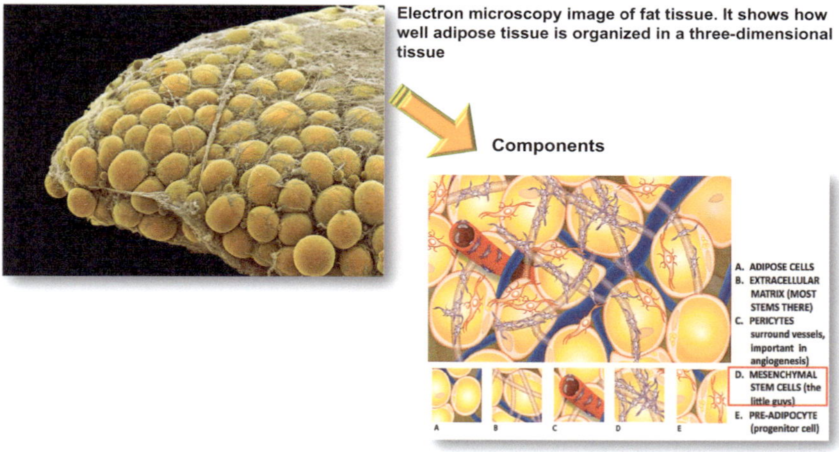

Fig. 1.5 Fat tissue as source of stem cells. A 3D electron microscopy image of fat tissue and how it is organized. Its components are described in detail

biomarkers, will be discussed. To understand their relevance and potentiality in this field, but also the related limitations (discussed in subsequent chapters), their biological features are discussed in the next paragraph.

1.2 Endothelial Progenitor Cells: Other Emerging Candidates for RegMed Therapeutic Treatment

The term endothelial progenitor cells (EPCs) was coined by Asahara et al. 1997 (the fathers of their discovery) for the following reasons: (1) to refer to circulating cells with surface antigens similar to those expressed by endothelial cells *in vitro*; (2) to be able to circulate and lodge in areas of ischemia or vascular injury; and (3) to facilitate the repair of damaged blood vessels or augment the development of new vessels as needed by a tissue (Asahara et al. 1997). EPCs have the capacity to proliferate, migrate, and differentiate into mature endothelial cells (ECs). Such functions were particularly underlined in 2004 by Urbich and Dimmeler (2004). In particular, they found EPCs to have three principal features: (1) to be non-endothelial cells, but to have the capacity to give rise ECs; (2) to show clonal ability to multiply; and (3) and stemness characteristics.

EPCs derived from HSCs in BM that represent, respectively, their principal origin and source (see Fig. 1.6), as discovered by Asahara et al. (1997), and subsequently confirmed by experimental studies on animal models, which have underlined the BM cell contribution in adult neovascularisation (Wojakowski et al. 2012; Pelosi et al. 2014). As a consequence, EPCs have been defined as CD34+ or

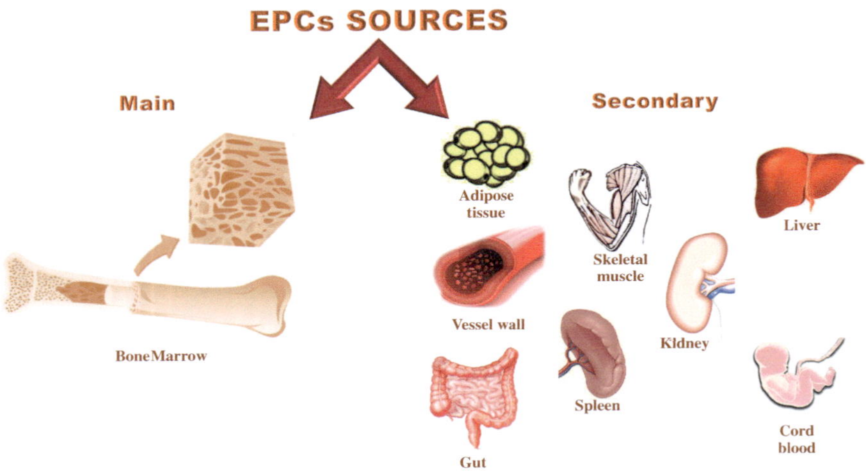

Fig. 1.6 Sources of endothelial progenitors. As illustrated, bone marrow is the principal but not exclusive source of these cells. Other sources have indeed been identified, including several organs (e.g.,. liver, gut and kidney) and tissues

CD133+ cells. In BM, EPCs live within BM stem cell niches, and an induced mobilization determines their release (as described below).

Recently, it has been found that other BM stem cells can also generate EPCs, including BM-myeloid cells and BM-derived mesenchymal stem cells (MSCs). BM-derived myeloid cells derive from HSCs and can be mobilized from BM. It has been detected that $CD14^{+}/CD34^{-}$ myeloid cells can co-express endothelial markers and form tube-like structures *ex vivo* (Pelosi et al. 2014). Thus, circulating BM-myeloid cells can differentiate into endothelial lineage, but with a lower proliferative capacity than HSCs or cord blood derived EPCs (see below). BM-MSCs can also generate mature ECs and contribute to neovascolarization. Circulating levels of these cells have been observed *in vitro* studies, and particularly in ischemic conditions. This has led to the hypothesis that, in the case of ischemia, there is also an active mobilization and contribution of BM-MSCs to endogenous cardiovascular repair (Pelosi et al. 2014).

Other origins and sources have been recently identified. Stem cell populations from adipose tissue, blood vessel wall, liver, intestine, spleen and kidney represent other potential candidates (see Fig. 1.6) (Pelosi et al. 2014). Among these, adipose tissue represents an alternative source of autologous adult stem cells (as previously reported) (see Fig. 1.5), which can be obtained in large quantities under local anesthesia and with minimal discomfort. Human lipoaspirate contains stem cells able to differentiate into several lineages. Furthermore, it has also been observed that isolated-tissue-derived, cultured, stromal-vascular $CD34^{-}CD31^{-}$ cell fractions can differentiate into ECs and promote angiogenesis (Watt et al. 2013; Murray et al. 2014). Adipose tissue is also a source of MSCs (see Fig. 1.5), originally identified in BM. MSCs can also differentiate into mature ECs in an appropriate micro-environment. In addition, they are characterized by their ability to modulate immune responses. This has led to them being considered as the best candidates for RegMed applications. Allogeneic transplants of these cells do not carry a substantial risk of immune rejection. This is linked to the capacity of MSCs to secrete various immunomodulatory molecules (as previously mentioned), which provide a regenerative microenvironment for a variety of injured tissues or organs, by limiting the damage and increasing self-regulated tissue regeneration (Watt et al. 2013; Murray et al. 2014). Autologous/allogeneic MSCs delivered via the bloodstream can contribute to the tissue repair process (Watt et al. 2013). Another origin and source of MSCs, recently identified, is the perivascular site, where they reside as pericytes or adventitial cells (see Fig. 1.5). This finding has generated important momentum in the field of adult stem cell research, by providing new insights on peri-vasculature. It represents an MSC niche *in vivo*, where local cues coordinate the transition to progenitor and mature cell phenotypes. Here, MSCs can stabilize blood vessels and contribute to tissue and immune system homeostasis under physiological conditions and assume a more active role in tissue repair in response to injury. The establishment of a perivascular compartment as the MSC niche provides a basis for the rational design of additional *in vivo* therapeutic approaches (Murray et al. 2014).

Another rich EPC source is cord blood (see Fig. 1.6). Cord blood contains higher numbers of CD133+ and CD34+ cells compared with peripheral blood from adults. In addition, a higher proliferative capacity and high levels of telomerase have been found in cord-blood derived EPCs. These characteristics are typical of stem cells and are very low or absent in other progenitor cell populations (Broxmeyer 2010).

This monograph is exclusively interested in EPCs derived from HSCs. The brief description of EPCs by other origins and sources is purely for educational purposes.

1.2.1 Evidence of Several Circulating Subpopulations of EPCs

EPCs are largely referred to as CD34+ CD133+ VEGFR2+ (or KDR+) cells. Recently, other progenitor populations have been considered in EPC studies. They include the following: circulating angiogenic cells (CACs), circulating endothelial cells (CECs), circulating H-progenitor cells (CPCs), and circulating endothelial progenitors (CEPs). They exhibit a common biological function in terms of playing important roles in tissue neovascularization, even if they have diverse characteristics. CAC and CEP cells represent variable proportions of CD14+ monocyte cells with different angiogenic properties. Despite their lower *in vitro* proliferation compared with HSCs or cord stem cells, they seem to have a similar ability to increase neovascularization, as reported in experimental models. This leads to the hypothesis that EPCs might essentially be H-monocyte-derived CD14+ cells with variable expression of CD34, CD133, CD45 and KDR and capability for angiogenesis (as aptly described in Balistreri et al. 2015a).

1.2.2 *EPCs: from BM progenitors to mature EC cells*

As previously reported, EPCs have the capacity to form new blood vessels. This requires multiple steps comprising EPC mobilization, migration, adhesion and differentiation (Li et al. 2012). The first step, the mobilization of EPCs from BM into the peripheral circulation, is crucial for their participation in postnatal vasculogenesis. To date, the mechanisms involved have not been entirely elucidated and remain under investigation. However, it has been demonstrated that, within the BM, EPCs are quiescent and tethered by integrins to stromal cells. In response to various cytokines and factors, EPCs can be converted into functional cells and released from the stem cell niche (Li et al. 2012). The principal authors of the release of these inductors are mature ECs under conditions of shear stress and hypoxia (Li et al. 2012). In particular, in the case of vascular occlusion, shear stress induces ECs to improve pro-oxidant enzyme expression, mediated principally by the most crucial transcription factor, the NF−kβ (Balistreri et al. 2015b). In the case of hypoxia, hypoxia-inducible transcription factor (HIF) and the levels of nitric oxide (NO) mediate the activation of several signaling pathways on ECs. As result, different growth factors, cytokines and chemokines are released, mediating EPC mobilization from BM (Li et al. 2012; Wojakowski et al. 2012) (see below).

Among the chemokines released by ECs, there are stromal cell derived factor 1 (SDF-1), angiopoietin (Ang)-1 and, probably the most important of all, vascular endothelial growth factor (VEGF) (Li et al. 2012). VEGF seems to determine a rapid EPC mobilization from BM, as discovered by Fox and colleagues in burned patients (Li et al. 2012). This led to the detection of the VEGF-(Receptor)-R-2 on EPCs, and also to define them as VEGFR2+ or KDR+, and CD309+.

After their homing, EPCs can release VEGF themselves and create a local angiogenetic environment. Recently, Li and colleagues reported that SDF-1 and VEGF mediate EPC mobilization, though their interaction with their respective receptors (leukocyte-derived seven transmembrane-domain receptor-CXCR4 and VEGFR2). This interaction determines the production of NO through the activation of eNOs. NO can stimulate metalloproteinase-9 (MMP-9), which results in the release of sKitL from the stromal cell membrane-bound kit ligand (mKitL). Proto-oncogene c-kit expressed by EPCs contributes to the retention of EPCs within the BM niches. C-Kit is also the receptor for sKitL and can be released from BM in response to binding to sKitL, resulting in the mobilization of c-Kit+ EPCs from the cell niche into circulation (Li et al. 2012).

Other factors such as erythropoietin (EPO) can mobilize EPCs (Povsic et al. 2013). Ang-1 seems to have a delayed and inhibitory effect, as evidenced in an unique study performed in 1999, where EPCs were defined as Tie+/Flk-1+/CD31 + cells (Ito et al. 1999).

The other crucial step is the adhesion of EPCs to an injured vessel wall. It occurs through the interaction of the glycoprotein ligand-1 (PSGL-1) expressed on EPCs with the P-selectin expressed on platelets, as suggested by Li et al. (2012). Platelets, within a few minutes of a vessel injury, aggregate on the exposed sub-endothelium. Adherent platelets express P-selectin on their surface and secrete high levels of SDF-1. In this process, circulating EPCs also upregulate PGSL-1 via stimulation of SDF-1, which interacts with the P-selectin ligand, thereby leading to EPC adhesion. Subsequently (within the next hours and days after endothelial disruption), apoptotic smooth muscle cells mainly contribute to SDF-1 release, which is required to sustain the process of vascular remodeling and repair (as summarized in Balistreri et al. 2015a).

The last step is the differentiation of EPCs in ECs. EPCs constitute a very reservoir of ECs, able to move into sites of injury, restore endothelium integrity and consent a physiological function (see Fig. 1.6). However, this function has been initially attributed to mature ECs. Indeed, ECs have been considered responsible for maintaining the homeostasis of the entire vessel wall (of arteries or veins), as well as for regenerating injured vessels (van Ierssel et al. 2014). In particular, it has been suggested that under the influence of paracrine mediators released from the injured segments and/or loss of contact inhibition, ECs migrate and proliferate (van Ierssel et al. 2014). Thus, the regenerative system has been conventionally established and identified in mature ECs adjacent to regions of injury (van Ierssel et al. 2014). Today, it is recognized that mature ECs possess limited regenerative capacity (Ma et al. 2015) and the discovery of EPCs has opened this question. The contribution of EPCs to vascularization has been demonstrated in animal models and in humans. Furthermore, it has been shown that a reduction in the numbers of circulating EPCs and/or alterations in their functions associated with different factors (see below) might have a remarkable impact on endothelium function and CVD onset and complications, and consequently in the survival of individuals with CVD (Ma et al. 2015).

1.3 Phenotypic Markers and Levels of Circulating EPCs in Adult Subjects

Current opinion states that the identification of circulating EPCs has to be based on a combination of specific and selective surface antigens, even if their expression is considered by some scientists to be insufficient and still missing. Indeed, they are recognized as CD34+ CD133+ VEGFR2+ cells. Thus, the use of the unique EPC pioneer marker, the CD34, which consented for the first time to Asahara and colleagues to discover EPCs (Resch et al. 2012), is not appropriate. It is usually expressed in both HSCs and mature ECs. CD34 is a 105–120 kD transmembrane cell surface glycoprotein, which is selectively expressed within the human and murine hematopoietic systems on stem and progenitor cells. It is also expressed in vascular endothelial cells. CD133, a more immature HSC marker protein, is a membrane glycoprotein, whose expression is related to differentiation of HSCs into EPCs. The third marker proposed is VEGFR2, also known as kinase insert domain receptor (KDR) or Flk-1, or CD309. It is mainly expressed on EC cells, and, in addition to EPCs, in low numbers on osteoblasts, pancreatic duct cells, neuronal cells, and lung epithelial cells, even if the biological role in non-endothelial cells remains unclear. VEGFR2 has been shown to be a vital promoter of pathological vascularization, including cancer and diabetic retinopathy, making it a critical target in the therapy of these diseases (Resch et al. 2012).

Other markers may be used for identifying circulating EPCs. These include platelet endothelial cell adhesion molecule-1 (CD31), CD146, von Willebrand factor (vWF), endothelial nitric oxide synthase (eNos), and E-selectin (Resch et al. 2012). C-kit and CXCR4 are the two further molecules commonly expressed on the surface of BM-derived EPCs. CXCR4, also known as fusion or leukocyte-derived seven transmembrane-domain receptor (LESTR), represents the receptor of SDF-1, which is highly expressed on the surface of CD34 positive cells. C-kit is a 145 kD transmembrane glycoprotein designed as CD117. This tyrosine kinase receptor and its ligand, stem cell factor (SDF), mediate pleiotropic functions, including cell survival, differentiation, homing, migration and proliferation, as well as functional activation. It is present on the surface of mast cells and cells of the erythropoid lineage as well as on multipotent stem and progenitor cells and megacaryocytes (Resch et al. 2012).

For identifying EPCs in cultures, a different expression profile surface markers has to be used. It supports the schema of human angioblast differentiation into EPCs proposed by Sieveking et al. (2008) (Fig. 1.7). EPCs are divided into early EPCs (eEPCs) and late EPCs (outgrowth endothelial cells, OECs), which originate by BM-hemangioblasts and are designed as CD133+ CD34+ and CD309+ angioblasts. CD133+, CD34+ and CD309+ angioblasts differentiate into early EPCs positive with high intensity for CD31, CD34, CD309 markers than the CD133, expressed with minor intensity. They give rise to late EPCs, having not only a high expression of CD34, CD309, CD31, but also of vWF, E-selectin, VE-cadherin and eNOs, which are also expressed in mature cells. Among the markers used in EPC cultures, the ulex europeaus lectin (UEA-I) has been also used for identifying EPC populations derived from mononuclear cell cultures (Resch et al. 2012). Clear differences in the nature and function have been shown between eEPCs and OECs. eEPCs seem to act in a paracrine manner, while the OECs directly participate in the formation of vascular-like structures (Medina et al. 2010). Certainly, to confirm these differences, a molecular analysis is necessary. Accordingly, a recent study, in order to establish a detailed molecular fingerprint of the two cell types, used a transcriptomic and proteomic analysis (Medina et al. 2010). The data obtained demonstrated that eEPCs are hematopoietic cells with a molecular phenotype linked to monocytes. Indeed, they mainly express specific hematopoietic (including RUNX1, WAS, PTPN6, HCLS1) and monocyte-like inflammatory (TLRs, CD14, HLR-DRA, lysozyme) transcripts. By contrast, the OECs exhibit a committed molecular phenotype linked to endothelial lineage. High levels of Tie2, eNos, Ephrin, caveolin-1, VE-cadherin, vWF have been assessed. This evidence supports the idea that OECs are progenitors committed to endothelial lineage (Medina et al. 2010).

Circulating EPCs, generally defined as CD34+ CD133+ VEGFR2+ cells, are characterized as existing in low levels in healthy adult subjects. Specifically, they represent only 0.0001–0.01% of peripheral blood mononuclear cells (PBMCs). However, it is now well recognized that both levels and viability of EPCs may be influenced by different factors, including methodological approaches (i.e., the timing and ways of taking samples), detection methods and their protocols, panel of antibodies used for their phenotypical evaluation, age of patients and their clinical conditions, and ethnicity of populations studied (Van Craenenbroeck et al. 2013) (see below).

Concerning the methods for detecting EPCs, several techniques have been proposed for isolating and quantitatively/qualitatively evaluating EPCs. They can first be summarized into two principal types: (1) immunohistochemistry or immunocytochemistry for quantifying tumor tissue EPCs (Gersh et al. 2009); or (2) methodologies for detecting EPCs in peripheral blood (Fig. 1.8). The latter are represented by four different methods available after the isolation of PBMCs:

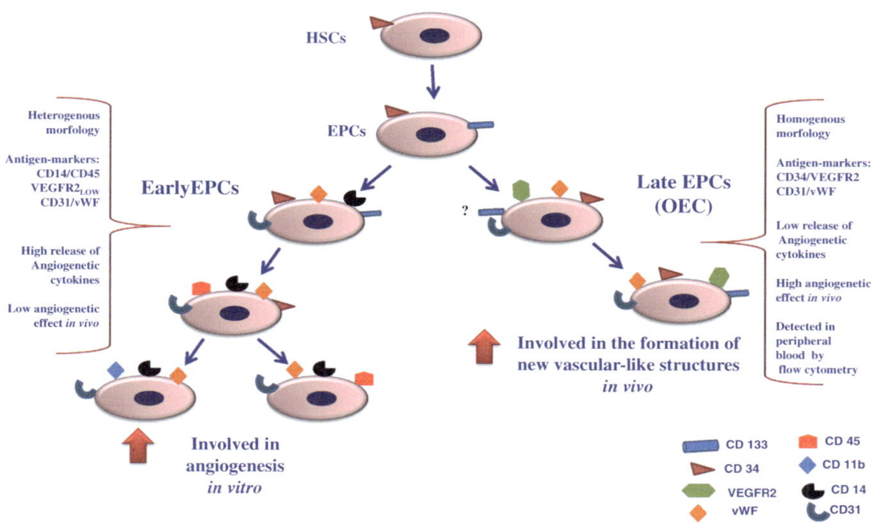

Fig. 1.7 Sieveking and colleague Model for EPC Differentiation As illustrated, CD133+ CD34 + and VEGFR2+ (CD309+) angioblasts give rise to early EPCs expressing high intensity CD31, CD34, CD309 markers, which differentiate in late outgrowth endothelial cells (OECs), having not only high expression of CD34, CD309, CD31, but also of vWF, E-selectin, VE-cadherin and eNOs

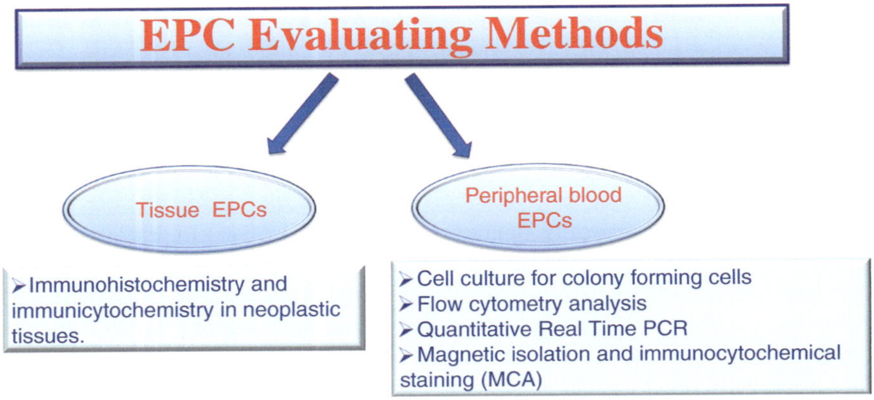

Fig. 1.8 Methods of evaluating EPCs. Two principal methods are used for evaluating the number and functions of EPCs. They are: immunohistochemistry or immunocytochemistry for quantifying tumor tissue EPCs and methodologies for detecting EPC in peripheral blood. These last comprise four different methods for detecting EPCs in PBMCs: *1* cell culture of colony forming cells; *2* phenotypic identification and enumeration of EPCs by flow cytometry analysis; *3* quantitative real time PCR; and *4* MCA method

(1) cell culture of colony forming cell unities (CFU), which consents to reveal EPC indicative features: high proliferative potential, expression of endothelial markers, endothelial morphology and formation of blood vessels in co-culture experiments (Gersh et al. 2009); (2) phenotypic identification and enumeration of EPCs by flow cytometry analysis according to the Duda protocol published in 2007 (Duda et al. 2007); (3) quantitative real time PCR, which permits detecting and quantifying EPC specific markers in pre-enriched PBMC cell population (Gersh et al. 2009); and (4) MCA method, which includes magnetic (M) isolation of CD34+ cells from PBMCs, followed by a CD133+ immunocytochemical (CA) staining (Gersh et al. 2009) (Fig. 1.8). To date, flow-cytometry and CFU assays are the two most used methods for EPC enumeration (Gersh et al. 2009).

It is important to stress that cell culture based assays are instrumental in the qualitative assessment of the different vasculogenic cell types, which is at least as valuable as numerical evaluation. In addition, a concordance between CFU assays and flow cytometric enumeration of CD34+/KDR+ cells is lacking, underlying again the different natures of the two cell types. From this point of view, flow cytometry analysis also allows EPC quality to be evaluated (Van Craenenbroeck et al. 2013). In the subsequent paragraph, EPC quantification using flow cytometry is described, along with its limitations.

1.3.1 *EPC Quantification Using Flow Cytometry and Its Limitation*

In human studies, multiparametric flow cytometry is the best method for performing quantitative and qualitative evaluations of EPCs. This consists of identifying a combination of antigens, for both the progenitor and endothelial phenotype. Based on the findings from Asahara et al. (1997), CD34 and KDR were postulated as being antigens that could characterize circulating EPCs. However, the CD34 antigen is not specific and is also expressed by mature endothelial cells (as previously reported). In an attempt to distinguish EPCs from CECs, CD133, as a marker of immature stem cells, was proposed as a third marker. However, more than 99% of CD34+/KDR+/CD133+ triple positive cells also express CD45, which is a pan-leukocyte marker. Since the endothelial lineage is negative for the CD45 antigen, even at the mRNA level, it is not surprising that several investigators observed that EPCs are not derived from CD45+ cells. As such, CD45 expression on putative EPCs became a bone of contention. At the same time, the technique of flow cytometry has evolved and a higher sensitivity is obtained for rare events and dim surface expression, allowing for improved distinction between CD45 negative, dim and bright positive events. In fact, CD34+/KDR+ EPCs dimly express CD45, adding a new dimension to the discussion. The identification of CD45 as a marker to be excluded opened perspectives and improved the precision of flow cytometric EPC detection, due to the creation of a more specific dump channel than the use of a lineage mixture (Van Craenenbroeck et al. 2013).

Based on a critical review of available studies looking at different EPC phenotypes as disease biomarkers, the CD34+/KDR+/CD45dim phenotype appears to be the best compromise in terms of sensitivity, specificity and reliability to quantify EPC in the clinical setting. The major drawback is the fact that CD34+/KDR+/CD45dim cells only constitute between 0.0001 and 0.01% of PBMCs, making their flow cytometric quantification a technical challenge. Rare event analysis involves background noise, which might lead to false-positive results. Consequently, signal enhancement and noise reduction are critical. Hence, several pre-analytical, analytical and analysis factors have to be taken into account in order to obtain reliable and reproducible quantitative and qualitative EPC evaluations (Van Craenenbroeck et al. 2013).

1.3.2 *Factors Modulating EPC Quantitative and Qualitative Evaluation by Flow Cytometry*

Several pre-analytical and analytical factors and analysis may affect the quantitative and qualitative EPC evaluations (Van Craenenbroeck et al. 2013). Concerning pre-analytical factors, it has been evidenced that *the type of blood sample*, including *whole blood or PBMCs* isolated by density gradient centrifugation, may affect evaluation. A remarkable discrepancy has been observed between the two approaches, with higher numbers in whole blood samples. This seems to be due to a selective loss of EPC in the progenitor cell compartment after density gradient centrifugation, or, most likely, caused by a higher number of false positive events in the whole blood sample. Indeed, from a statistical point of view, the accuracy is higher when measuring in an enriched mononuclear cell fraction since more events of interest can be recorded compared with a lysed whole blood sample. As a consequence, the standard error will be lower. Furthermore, CD34+ cells can be isolated using magnetically labeled beads prior to flow cytometry. Combined results from clinical trials indicate that the median yield and purity following magnetic bead isolation of CD34+ cells are 71 and 85%, respectively. CD34+ cell enrichment increases the cell population of interest (i.e., CD34+/KDR+/CD45dim cells,) which reduces the coefficient of variation of measurements. Other pre-analytic limitations include *blood collection tube and handling temperature*. The first tube to be filled is a waste tube to prevent contamination with circulating endothelial cells, which could be released during the venipuncture. The blood sample for flow cytometric enumeration is collected in tubes containing an anticoagulant, such as ethylene-diamine-tetra-acetic acid [EDTA], heparin, or acid citrate dextrose [ACD]. It has been demonstrated that these three anticoagulants on lymphocyte immunophenotyping have similar effects (Van Craenenbroeck et al. 2013). To be specific, enumeration of lymphocyte subsets is comparable between fresh and stored samples for 24 or 48 h and using ACD and heparin, but not when using EDTA. However, reports on the effect of different anticoagulants on the specific identification of CD34+/KDR+ cells are lacking. Concerning *storage time*, it has been noted that the stability of surface markers of CD31+/CD133+ cells significantly decrease after 24 h. However, sample storage time can be prolonged by the addition of Trans-Fix (Cytomark, England) stabilizer reagent to whole blood immediately after venipuncture. Indeed, Trans-Fix is known to be quite suitable for stabilizing the biological integrity of lymphocytes, and, to a lesser extent, also that of the monocytic cell population. However, recent findings have shown that Trans-Fix can be reliably used for the analysis of CD34+ cells and for up to 7 days after sample treatment. *Handling temperature* is another factor to take into account, since processing samples at room temperature increases fluorescence intensity in leukocyte phenotyping compared with working at 4 °C. However, handling temperature and possible differences between ACD and heparin on a small number of antigens have been evaluated. A profound increase in fluorescence intensity for samples stored at room temperature as soon as 6 h after sampling has been observed, which was not observed in samples stored at 4 °C, even after 24 h. This increase is higher in heparin samples. In summary, ACD is preferred for the anticoagulation of blood samples; staining of samples should be performed at 4 °C; and fresh blood samples should be processed as soon as possible, with a maximum storage time of 6 h in order to guarantee optimum sample condition (Van Craenenbroeck et al. 2013).

Other limiting factors are the c*onditions for sample collection and subject-related confounding factors* (Van Craenenbroeck et al. 2013). Diurnal variations characterize circulating EPC levels with a significant evening peak. Due to this repeated blood samples should be planned at a similar time of day. In addition, as reported in Chap. 2, physical exercise influences the number of EPCs. Following one maximal exercise bout, numbers of circulating CD34+/KDR+ cells increased in both healthy subjects and in patients with CVDs (see Chap. 2). In healthy subjects at least, this effect was observed until 24 h after exercise and the numbers of CD34 +/KDR+ EPCs had returned to baseline after 48 h. Therefore, patients should be asked not to engage in *vigorous physical activity* at least 24 h prior to blood sampling. In addition, *specific diets and several dietary components* could affect the number of EPCs. Cocoa flavonoids, for example, have received a lot of interest in recent years as they substantially reduce the risk of cardiometabolic disorders. The possible mechanisms for this risk reduction include the release of EPCs after chronic ingestion. Red wine and green tea are known for their beneficial effects on endothelial function, which is associated with a rapid increase in EPC levels. Therefore, to exclude a possible effect of any known or unknown dietary components, samples should be taken after an overnight fast. *Numerous drugs* are known for influencing numbers of EPCs, as summarized in Chap. 2. Finally, several investigators have demonstrated a deleterious effect of both *active and passive smoking* on CD34+/KDR+ EPC numbers. It is therefore recommended to instruct patients to undergo smoking cessation 6 h prior to blood sampling (Van Craenenbroeck et al. 2013).

Among the factors affecting EPC enumeration, *data acquisition* is crucial and is influenced by the type of flow cytometry used, the research protocol applied, as well as other factors. A *flow cytometer is* a multicolor flow cytometry system capable of detecting at least five parameters simultaneously as a minimum requirement. Obviously, the performance of the flow cytometer has to be checked daily with standardized beads. Every laboratory should define its own acceptability criteria regarding the magnitude of the instrument variation. Another factor is *depletion of erythrocytes and wash/no wash protocols:* some researchers advocate a no-lyse protocol. However, depletion of erythrocytes remains useful in EPC analysis since it enables more careful data acquisition by eliminating a substantial amount of irrelevant red blood cells. Wash steps definitely lead to systematic cell loss, but no-washing decreases the signal-to-noise ratio and thus requires caution. Erythrocyte lysis without washing is typically performed when absolute cell counts are commanded. For this purpose, the "guideline for the flow cytometric enumeration of CD34+ hematopoietic stem cells" recommends a no-wash method and the use of a fixative-free ammonium chloride (NH_4Cl)-based erythrocyte lysis reagent. The cell loss associated with erythrocyte lysis has been shown reduced in case of a fixative-free lysing reagent. An appropriate reagent to lyse erythrocytes is NH_4Cl, which also contains EDTA, thereby profoundly reducing the stickiness of the cells and thus improving pelleting. *Increasing the signal to noise ratio* is another factor that affects EPC enumeration: rare event analysis has to deal with background noise, which may result in false-positive events. As a consequence, signal enhancement and noise reduction are critical (e.g., to prevent unspecific binding of antibodies and auto-fluorescence). In view of the Poisson distribution of rare events, it is mandatory to acquire a very large amount of total events (1–2 million) in order to gather a representative number of CD34+/KDR+/CD45 dim events of interest. Increasing cell concentration and increasing throughput by fluorescence threshold triggering, are valuable means, whereas increasing the sample flow rate is not an option, since this will broaden the sample core stream, thereby increasing the coefficient of variation. In summary, for CD34+/KDR+/CD45dim EPC analysis, the following aspects require attention. Rare events should be preferentially labeled with the brightest and tightest fluorochromes. Conjugated antibodies should be used whenever possible. Indirect staining often produces an exceedingly high proportion of false positives in rare event detection. Potential non-specific binding of antibodies to non-relevant cells can be minimized by pre-incubation of the cells with a 'blocking' serum before the actual staining with antibodies for EPC enumeration. Dead cells are a major source of non-specific staining, so the use of a viability stain, such as 7-aminoactinomycin D (7-AAD), in each test tube is certainly advocated. The addition of a nucleic acid stain is meaningful because it facilitates the exclusion of unlysed erythrocytes, platelets and debris from the analysis. Before starting flow cytometry, extensive instrument cleaning and washing are necessary to remove all residual cells and debris. Large cells and clumps of cells expose stronger autofluorescence, and can be averted by gating out aggregates based on pulse shape analysis or doublet gating (FSC area/FSC height plot). *Controls to evaluate the background in rare event analysis* also impact EPC enumeration. The use of advanced multi-parameter gating strategies has de facto replaced the necessity for controls. Still, depending on the gating design, fluorescence-minus-one (FMO) controls, wherein all of the test antigens are included except for the antigen of interest, helps in delineating the threshold for positivity. In particular, FMO controls for KDR and CD34 are valuable in reducing subjectivity in analysis and in preventing gating out false-negative events in dim regions. In addition, an 'unstained cell sample', i.e. a sample processed with the identical protocol but without the addition of antibodies, may be of use when measured as a negative control to inspect the degree of autofluorescence in the cell sample. The use of isotype control tubes as a measure of non-specific binding is not encouraged. Specifically, variations in the staining of the same cell populations by different isotype controls have been reported. In this respect, differences in protein concentration and in fluorescence to protein (F/P) ratio between the antibody of interest and the isotype control may obscure the reliability of the rare event analysis. On the other hand, even at the same F/P ratio and at the same AB concentration, conjugates might still act in a significant different manner. Finally, *data acquisition* is a factor. The forward scatter (FSC) detector is commonly chosen to define the threshold for the detection of cells of interest. The lower threshold of FSC is set so that all lymphocytes are included. Sometimes, a certain amount of platelets is present, which might be well delineated in terms of FSC, or gated out based on the FSC/nuclear stain plot. The population of counted events is defined electronically during the analysis and is often set as a gate at the FSC/SSC plot or the SSC/CD45 plot. It is important that sufficient numbers are counted to produce a statistically valid result. Reducing the number of events analyzed per test tube reduces the reliability of the EPC estimation (Van Craenenbroeck et al. 2013).

Data analysis represents another affecting factor. Among other things, it is modulated by a *gating strategy*: it has been demonstrated that the use of sequential gating, FMO controls, automated compensation and biexponential display leads to increased accuracy of EPC quantification and identification of new subpopulations of EPC. However, Schmidt-Lucke and colleagues recently applied a modified International Society of Hematotherapy and Graft Engineering (ISHAGE) protocol to CD34+/KDR+/CD45dim EPC enumeration. This protocol has been introduced in order to facilitate inter-laboratory comparison for hematopoietic stem cell enumeration, avoiding isotype controls and using sequential gating, the dim expression of CD45 and the low side scatter of stem cells. The ISHAGE protocol gave the most reproducible results for CD34+ stem cell enumeration between centers. Actually, studies comparing this new gating strategy with the use of Khan's recommendations for rare event analysis are lacking, i.e. the use of a viability stain and nuclear stain together with a dump channel. Data analysis is also modulated by *expression of results*: data can be expressed as the absolute number of EPC per volume of blood, either by combining a separate cell count from a hematology analyzer with flow cytometric population data or by adding an internal microsphere counting standard to the flow cytometric sample. For example, in the ISHAGE protocol, the number of CD34+ cells is expressed as a percentage of CD45+ cells and then multiplied by the available white blood cell count. This method has several drawbacks because of the magnification of the measurement error, not accounting for hemodilution, hemoconcentration or the mobilization of white blood cells. Therefore, EPCs are now preferentially expressed as a percentage of the counted mononuclear cells after exclusion of doublets, dead cells or un-nucleated cells/debris. This approach can be used in the lysed whole blood technique (percentage of single, nucleated mononuclear fraction) as well as after PBMC isolation (percentage of single, alive and/or nucleated mononuclear fraction). For quantification after CD34+ cell enrichment, it suggests an expression in terms of the percentage of counted CD34+ cells (single, alive and/or nucleated) in order to minimize the effect of the isolation procedure. In addition, *statistical considerations* have a bearing: the amount of events that needs to be recorded in order to obtain a reliable result is extremely important when dealing with rare events. As previously stated, acquisition of a larger number of positive events leads to a lower standard error (SEM = $\sqrt{\text{positive events}}$). This is one of the reasons to focus on CD34+/KDR +/CD45dim cells. The addition of a third marker (e.g., CD133) significantly decreases the number of events. According to the modified ISHAGE protocol, at least 100 CD34+ events should be recorded. Furthermore, repeated measurements for the determination of the coefficient of variation are mandatory in order to tag aberrant results (Van Craenenbroeck et al. 2013).

1.3.3 Pathological or Physiological Conditions as Factors Able to Modulate EPC Quantitative and Qualitative Evaluation

A number of factors need to be taken into consideration in performing EPC quantitative evaluations and in their interpretations. In case of particular patients, such as intensive care patients, several affecting factors have to be considered. Firstly, the site of blood sampling (e.g., central venous linear arterial catheter) may be important, depending on specific homing of EPCs. Therapeutic interventions can influence EPC enumeration; for example, the number of EPC increases during haemodialysis. In addition, pharmacological compounds are often interrupted during particular cares or clinical interventions. This may determine the possibility of an acute change in EPC number. Furthermore, patients with chronic kidney diseases have different confounding factors, which might modulate EPC quantitative evaluations. For example, the administration of recombinant human erythropoietin (rh-EPO) or erythropoietin stimulating agents (ESA) as a treatment of renal anaemia, has been shown to be a strong stimulus for EPC mobilization and proliferation *in vitro* and *in vivo*. Dialysis may influence EPC numbers and function by removing the uremic toxins, which are known to be harmful for EPC biology (as summarized in Balistreri et al. 2015a).

Another particular factor to consider in EPC quantitative investigations is age. In children and infants, EPC quantification may be very important in providing a minimally invasive way to study cardiovascular risk. However, some recommendations are essential for performing these evaluations. Firstly, EPC numbers are higher in children than in adults, although more research is needed to study the effects of puberty and growth. This decline in EPC numbers with age could be caused by an impaired proliferative capacity of BM stem cells as described in Chap. 2. In performing studies on children and infants, careful consideration also is necessary to keep blood volumes to a minimum. It has been observed that children with acute mild infection had a significant impairment of endothelial function as compared with healthy controls. Based on these observations, EPC quantification in children suffering from an acute infection is not recommended and should preferably be postponed for at least two weeks (Van Craenenbroeck et al. 2013).

In the next section, in order to elucidate how all of these factors can influence the number and functions of EPCs, the complex process of regulation of EPC biology and function will be described.

1.4 The Complex Process of Regulation of EPC Biology and Function: New Insights on the Key Role of Notch, ROS and Wnt Pathways and Their Crosstalk, and Epigenetics

As previously mentioned, pathological or physiological factors can affect the circulating levels of EPCs and their functionality. To understand how the biological function of EPCs is altered, it is important to note that a complex control process is involved with a large number of regulators. However, the actors involved have been yet all been completely detected. Their identification might allow targets (agonists and/or antagonists) to be developed for the treatment of several pathologies.

However, a very recent genome-wide epigenetic and proteomic analysis revealed that EPC dysfunction is significantly linked to an altered *Notch signaling* (Karcher et al. 2015). In particular, it has been observed that a high-salt diet decreased both the number and impaired the function of EPCs derived from BM. This effect has been evaluated using Sprague–Dawley rats placed on a normal-salt diet (NSD, 0.4% NaCl Dyets) or a high-salt diet (HSD, 4% NaCl Dyets) for 7 days. Following 7 days on either diet, the rats were killed with Beuthanasia-D Special (Merck) and both femurs and tibias were surgically dissected. The BM mononuclear cell population was isolated and characterized. Thus, it has been shown that a HSD (4% NaCl) caused a decrease in the relative number of EPCs in the BM mononuclear cell fraction but it did not alter the cellular marker expression of isolated and 14-day cultured EPCs. However, following *in vivo* transplantation they were unable to determine angiogenesis. By contrast, *in vivo* transplantation NSD EPCs demonstrated a robust therapeutic restoration of stimulated angiogenesis. Interestingly, this also revealed that the effect of dietary salt persisted through 14 days of cell culture. Once isolated, all cells underwent the same culture and purification protocol with the same conditions, demonstrating that these cells retain a "memory" of their dietary salt in *vivo*. During the 14-day cell culture preparation of the EPCs, the cells went through many rounds of doubling suggesting epigenetic changes *in vivo* impacted the cellular phenotype *in vitro*. Among epigenetic changes, DNA methylation plays an important role in the gene expression. Thus, the rate of methylation from NSD and HSD EPCs has been evaluated. The same global rate of methylation and overall number of methylated regions have been observed between NSD and HSD EPC. The differences in methylation were confined to individual genes where the methylation rates of CpG islands in transcription start sites, intragenic and intergenic regions varied between NSD and HSD EPCs. In particular, neurogenic locus notch homolog protein 4 (Notch 4) was found to have increased methylation in HSD EPCs and had suppressed protein expression in HSD EPCs. Thus, it has been hypothesized that suppressed Notch in HSD EPCs may account for the inability of HSD EPCs to improve angiogenesis to the same degree as NSD EPCs. Other Notch genes including Notch1, Notch2, and Notch3 were evaluated in this proteomic and DNA methylation analysis; however, none of the other family members met the strict criteria to be a potential candidate (Karcher et al. 2015).

For a clearer understanding of these data, a brief description of this signaling pathway and its relevance in vascular biology and particularly in EPC biology follows

- ***Notch Signaling and Vascular Biology***: *The Notch signaling pathway is evolutionally conserved and regulates the cell fate decision during many developmental processes. It consists of four members (Notch* 1, 2, 3, *and* 4) *and five ligands (Delta-like ligands* 1, 3, *and* 4 *and Jagged* 1 *and* 2). *Both receptors and ligands are membrane-spanning proteins. Ligand binding induces a conformational change that allows the first proteolytic cut by A Disintegrin and Metalloprotease (ADAM) which removes the extracellular portion of Notch and creates a membrane-tethered intermediate that is a substrate for* $\gamma-$secretase, *a cleaving protease complex.* $\gamma-$secretase *in turn generates the active form of Notch (Notch intracellular fragment, NIC) which translocates to the nucleus where it binds the transcriptional factor CSL (CBF1, Suppressor of Hairless, Lag-*1) *also known as recombinant signal binding protein 1 for* $J\kappa(RBP-J\kappa)$. *Such NIC binding displaces repressor molecules and promotes the recruitment of coactivator molecules. This in turn activates the transcription of specific Notch target genes such as Hes (hairy/enhancer of split), Hey (Hes-related proteins), Nrarp (Notch-regulated andrin repeat protein), cMyc, cyclin D*1, *and many other genes that control EPC and EC proliferation, differentiation and apoptosis, as well as stem cells maintenance and angiogenesis. For instance, it was shown that perturbation of the Notch signaling pathway leads to developmental abnormalities of the vascular system. In human vascular diseases, haploinsufficiency of the Jagged-*1 *(Jag-*1) *gene, a ligand of the Notch receptor, is closely related to the dominantly inherited Alagille syndrome, which is characterized by various pleiotropic developmental problems with accompanying features of congenital heart defects and cardiovascular anomalies. A mutation in the Notch-*3 *gene leads to a severe human degenerative vascular disease termed cerebral autosomal dominant arteriopathy with subcortical infarcts and leukoencephalopathy. Genetic mutation analysis of mice lacking Notch receptors or their ligands revealed abnormalities during blood vessel formation, arteriovenous specification and impaired proliferation and migration of endothelial cells. Loss of Notch-*1, *Notch-*4, *Notch-*3, *Delta-like ligand (Dll)-*4, *or Jag-*1 *lead to defects in angiogenesis and vascular development. Likewise, the knockout of Notch transcriptional targets Hey*1 *and Hey*2 *impaired vascular development, with vessels still forming* de novo *but remaining small and failing to develop. Notably, the Notch ligand Dll-*4 *has been shown to be a pivotal regulator of tumor angiogenesis and vascular development, linking EC bioactivity to the tissue environment* via *Notch ligand receptor signaling. Notch ligand/receptor-mediated signaling may play a key role in the function and kinetics of BM-derived EPCs. It is speculated that EPC maintenance and their proliferation, differentiation, mobilization and subsequent contribution to* *vascular regeneration may be regulated by distinct and defined signals triggered and mediated by specific Notch receptor/ligand interactions between EPCs and the BM environment. Notch signaling may not only be involved in the interaction of EPCs with the BM-niche but may also play pivotal roles in the peripheral part of EPC biology by influencing and regulating homing, onsite differentiation and integration of EPCs into target tissues. It should be further noted that the eminent role of Notch signaling in tumor angiogenesis is paralleled by the critical impact of EPCs on tumor vascularization, growth and progression, indicating a possible regulatory role of Notch signaling on another peripheral function of EPCs* (Kwon et al. 2009; Balistreri et al. 2016).

Similar to the data on high-salt diets, another recent report evidenced that uncontrolled hyperglycemia determines not only oxidative stress (ROS) and Advanced Glycation End products (AGE) accumulation, but also epigenetic changes with long-term effects, such as histone modifications, DNA methylation and microRNA-mediated translational control, which appear to modulate synthesis of proteins in endothelial (and other) cells, such as EPCs (Rajasekar et al. 2015). Modulation of the Notch signaling pathway and EPC dysfunction has been associated with these changes related to uncontrolled hyperglycaemia and the supposed "glycemic memory". Epidemiological and experimental evidence accumulated over the last two decades has proven the existence of a glycemic memory, a phenomenon triggered by uncontrolled hyperglycemia in the early stages of disease. The consensus view is that an effective way to diminish diabetic vascular complications is early diagnosis and initiation of appropriate glycemic control measures. This will prevent the imprinting of glycemic memory in endothelial (and other) cells, such as EPCs, and consequently result in a slower progression of vascular complications. While oxidative stress and AGE accumulation seem to be the key biochemical drivers of glycemic memory, epigenetic changes such as histone modifications and DNA methylation appear to modulate glycemic memory at the molecular level. Furthermore, it seems likely that endogenous EPCs represent a significant cellular component where all these biochemical and molecular processes take place. Considering their extended lifespan and vascular homeostatic function, it is likely that glycemic memory affects EPCs and that this contributes to the impaired vasoreparative capacities of diabetic tissues (Rajasekar et al. 2015).

There is a pressing necessity for new therapeutic targets to address the glycemic memory phenomenon. Among the many candidates, drugs that modulate epigenetic mechanisms have the potential to "erase" glycemic memory and, consequently, delay progression of microvascular complications.

These recent findings reveal the key role of epigenetic mechanisms and the specific signaling pathways, such as the Notch pathway, as modulators or regulators of EPC biology. Recent growing evidence reports the existence of a strict crosstalk between reactive oxygen species (ROS), Notch and Wnt/β−catenin signaling pathways as key regulator of EPC biology and cardiovascular biology (Caliceti et al. 2014). It is well recognized that ROS have long been deemed as noxious molecules for tissues, organs and systems, such the cardiovascular and nervous systems and the pancreas, and as remarkable risk factors of CVDs (i.e. systemic and pulmonary hypertension, atherosclerosis, cardiac hypertrophy and heart failure), diabetes, cancer and dementia. With years of effort, ROS are becoming increasingly recognized as an important modulator for a variety of biological functions and pathophysiological states. Recent evidences suggest an even more significant role of ROS: Notch and Wnt/$\beta-$catenin signaling modulation. A clear distinction between Notch and Wnt responses is vital for appropriate and robust cell-fate decisions, and ROS modulation of these signaling pathways would provide clues for clinical strategies and drug discovery targeting the above-mentioned diseases, i.e. CVDs, pathways, diabetes, cancer and dementia. ROS modulation of Notch and Wnt signaling regulates vascular development in different aspects, including stem cells differentiation, angiogenesis, VEGF signaling, endothelial as well as cardiac progenitor cells recruitment, and vascular cell migration (Caliceti et al. 2014). Nonetheless, more details regarding ROS signaling and pathophysiological functions remain to be elucidated. A deeper insight into the mechanism of how ROS affect normal vascular development, especially cardiac progenitor cells, EPCs and progenitor of vascular smooth muscle cells, and their differentiation from stem cells, could contribute to a brighter future for RegMed in cardiovascular, cancer, diabetes and dementia therapies. Factors (i.e. probably epigenetic factors) that selectively control ROS modulation of Notch and Wnt/$\beta-$catenin signaling pathways could have therapeutic effects repressing angiogenesis in tumors or favoring it in ischemic tissues.

For a clearer understanding of these data, a description of: Wnt/$\beta-$catenin signaling pathway and its relevance in vascular biology and, in particular, in EPC biology as well as the epigenetic mechanisms are reported:

- *Wnt*/$\beta-$catenin ***signaling***. *Notch signaling modulates endothelial homeostasis by cross talking with other signaling pathways, such as receptor tyrosine kinases (i.e. VEGFR*2) *and estrogen receptor. Crosstalk between Notch and Wnt has also been described. The Wnt signaling pathway, also called Wnt*/$\beta-$catenin *signaling, plays a key role in vascular biology. Mice deficient for Wnt displayed vascular abnormalities including defective placental vasculature. Knock-out mice for the Wnt receptor gene, Frizzled*5, *died* in utero *due to defects in yolk sac angiogenesis. Defects of the* $\beta-$catenin *gene in ECs caused aberrant vascular patterning and increased vascular fragility. The canonical Wnt signaling pathway is driven by* $\beta-$catenin, *a scaffold protein, linking the cytoplasmic tail of classical cadherins in the endothelium* [*vascular endothelial* (V*E*) *cadherin and N-cadherin*] via $\alpha-$catenin *to the actin cytoskeleton. Without Wnt stimulation, cytoplasmic* $\beta-$catenin *levels are kept low by a degradation complex, consisting of Axin, Adenomatous polyposis coli* (*APC*), *Casein kinase*1*a* (*CK*1*a*), *and Glycogen synthase kinase* 3$-\beta$(GSK3$-\beta$). *Binding of Wnt to its receptors, Frizzled and lipoprotein receptor-related protein* (*LRP*), *leads to inhibition of the degradation complex function, enabling* $\beta-$catenin *signaling. Wnt allows* $\beta-$catenin *to accumulate and translocate to the nucleus where it* *binds to several transcription factors, for example, T-cell factor (TCF) and LEF-1. Dishevelled (Dvl) is an essential adaptor protein for Wnt signaling that interacts with several molecules, including Axin, inactivating the β−*catenin *degradation complex. Dvl has a dual role: it is an activator of downstream Wnt signaling and an inhibitor of Notch activity. Thus, Dv1 is a key regulator of cell-fate decisions in which Wnt and Notch have opposing effects. The non-canonical Wnt pathway can be categorized largely into two classes, the Wnt/Ca2+ and Wnt/planar cell polarity (PCP) pathways. The Wnt/Ca2+ pathway, mediated by G protein signaling, stimulates the release of intracellular Ca2+ and activation of Ca2+ -sensitive kinases, such as the protein kinase C and Ca2+ -calmodulin kinase II. PKC is a family of Ca2+ -dependent and Ca2+ -independent isoforms that have different distributions in various blood vessels, and individual members can have different roles in a plethora of biological and pathological events. The PCP pathway, also referred to as the Wnt/Jun-N-terminal kinase (JNK) pathway, was originally identified as a pathway affecting cytoskeletal reorganization. This pathway activates small GTPases, including RhoA, Rac, Cdc42, and JNK* (Caliceti et al. 2014).

- ***Epigenetic mechanisms and regulation of EPC biology***. *Epigenetic processes are described as changes in gene expression and phenotype caused by alterations in the genome that do not involve changes in the DNA sequence. Epigenetic mechanisms can be widely classified based on their short- or long-term effect. The short-term effect often involves a rapid response to varying environmental factors and is non-heritable in nature. The long-term effect on the other hand creates a persistent change that is stored as "memory" and passed on to the offspring, usually in response to excessive exposure to long-acting stimuli. However, it has been reported that even a transient change in the microenvironment can induce persisting epigenetic effects. Epigenetic mechanisms are divided into three main categories:* ***post-translational histone modifications (PTHMs), DNA methylation, and microRNA-mediated translational control***. ***PTHMs*** *comprise a complex variety of biochemical reactions such as acetylation, methylation, phosphorylation and ubiquitination. A combination of different PTHMs collectively acts in unison to create a local chromatin modification pattern. Histone acetylation of lysine residues in H3 and H4 subunits results in euchromatin conformation (active state). This reversible acetylation reaction is catalyzed by the enzyme histone acetyl transferase (HAT), while the deacetylation is induced by the enzyme histone deacetylase (HDAC). Deacetylation converts euchromatin to heterochromatin (inactive state) and, consequently, causes transcriptional silencing. Histone methylation is another important PTHM where the enzyme histone methyl transferase (HMT) adds methyl groups from S-adenosyl methionine to the amine terminal of lysine or arginine. Unlike histone acetylation, histone methylation activates or silences transcription based on the position of lysine or arginine residues and the number of methyl groups added. For example, mono-, di- and trimethylated H3K4 result in euchromatin, while mono-, di- and trimethylated H3K9, H3K27, and H4K20 result in heterochromatin.* ***DNA methylation*** *is characterized by the* *addition of a methyl group to the cytosine residue in the CpG dinucleotide sequence by DNA methyl transferase enzymes* (*DNMTs*). *The extent of DNA methylation at CpG islands* (*genomic regions with high density of CpG-dinucleotides*) *located in promoter regions determines the transcription status of the gene. DNMT1 classically functions as a maintenance DNMT since it conserves the methylation pattern during replication, while DNMT3A and DNMT3B create* de novo *methylation marks. DNA methylation at promoter regions is usually inversely associated with the regulation of gene expression. Therefore, hypermethylation is associated with a blockade of normal gene expression. DNA demethylation is another important phenomenon observed during embryogenesis and certain disease conditions. The demethylation process often results from the absence of methylation pattern maintenance by DNMT*1 (*passive demethylation*) *or enzyme-mediated excision of methylated cytosine* (*active demethylation*)*. A class of enzymes called demethylases belonging to the family of MBD* (*methyl CpG-bindingdomain*) *proteins mediates the active demethylation process. DNA methylation/demethylation is a dynamic process occurring simultaneously that regulates cellular functions and selectively triggers or represses gene expression.* ***MicroRNAs*** (***miRNAs***) *constitute a very different class of epigenetic controls. These single stranded noncoding RNA molecules interact with transcribed mRNA to inhibit translation, unlike PTHMs and DNA methylation, which act on the genome before transcription. miRNAs are transcribed independently and are present in the intergenic regions, exons, or protein coding introns. miRNAs bind to the* 3′ *untranslated region* (3′UTR) *of targeted mRNA and inhibit translation by either degradation of mRNA or destabilization by cleavage and deadenylation. MicroRNAs* (Caliceti et al. 2014).

- ***MicroRNAs*** *exist in a variety of cells. Some researchers have found a positive or negative correlation between levels of some microRNAs and cellular activities including EPCs isolated from patients with coronary artery disease* (*CAD*)*. MicroRNAs profiles of early EPCs, late EPCs and human umbilical vein endothelial cells have been identified. Notably,* 722 *microRNAs are expressed in the above three endothelial lineage cells, among which are* 38 *microRNAs are specifically expressed in late EPCs and* 30 *microRNAs are specifically expressed in early EPCs. These results indicate that microRNAs may play an important role in regulating EPC functions. Recent studies have suggested that microRNAs regulate the proliferation, mobilization, migration, differentiation, apoptosis, autophagy, senescence, adhesion and tubule structure formation of EPCs and EPC-induced angiogenesis* (Qu et al. 2015).

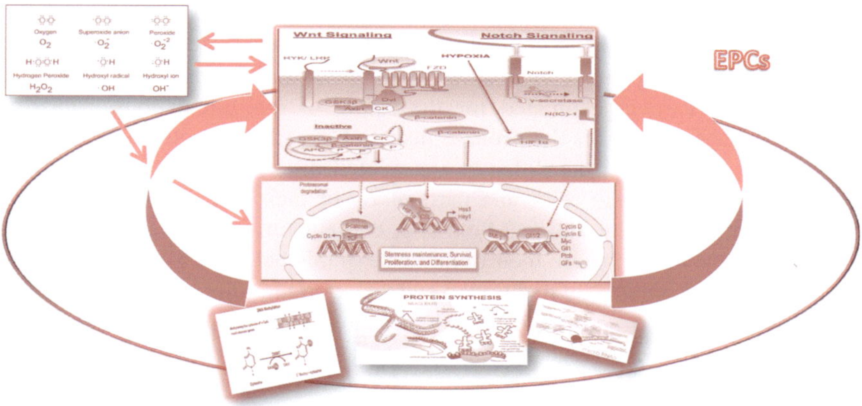

Fig. 1.9 Key regulators of EPC biology and EPC mediated regeneration. ROS, Notch and Wnt/β-catenin signaling pathways, and epigenetic factors modulate EPC biology and functions

1.5 Conclusions

BM-EPCs are emerging as potential candidates of cell-based therapy. They exhibit features, as described in this chapter, which contribute to the understanding of the relevance and potential of these cells in cell therapy, and, co-temporally, their potential as possible biomarkers for diseases (see Chap. 2 for more information). However, at the same time, some points have emerged that are critical of EPCs, including their limited number in circulation in adult humans; their lack of a definitive phenotype and characterization; and the capacity of diverse factors to modulate their number and function, from factors linked to analysis and methodology to those relating to physiological and pathological conditions. These limitations have led some researchers to unfavorably judge EPCs, which might lead the scientific community to pursue new investigations on other progenitor cells. However, the very limitation of these promising cells is the reduced knowledge of their complete biology, their characterization and function. Recently, some important advances have been obtained, including insights into the identification of several factors with the key role of regulators (i.e. ROS, Notch and Wnt/β-catenin signaling pathways, and epigenetic factors) of EPC biology and EPC-mediated regeneration (Fig. 1.9). Modulation of epigenetic factors and the Notch and Wnt/β-catenin signaling pathways could have therapeutic effects that are able to repress angiogenesis in tumors or to favor this last in chronic age-related diseases.

Thus, it is imperative to improve investigations on EPCs in order to re-examine the actual evidence on EPCs, by also evaluating all of their subpopulations, their functions and potentiality. Future investigations at several levels from genomic assays to proteomics and epigenomics analyses are necessary. Their results may provide clarity on the biological role or roles of several EPCs before their clinical application as both biomarkers and therapeutic agents.

The suggestion that, this report aims to put forward, is that the research and study on these cells should continue, because they might represent a force point for RegMed in the subsequent years in terms of counteracting the growing incidence of age-related diseases, as estimated for the future of our populations. Thus, the successive chapter describes their role as disease biomarkers and therapeutic agents in a general overview in the different age-related diseases, in order to provide a general overview of these cells, and particularly of their potentiality (with all the related limitations) in translational field and, consequently, in the clinical applications. The aim is to emphasize their importance in this field and to try to counteract the limitations (as previously mentioned and as reported in the successive chapter), by adopting common and standardized measures both in research and clinical applications, and other strategies to improve their therapeutic efficacy (see Chap. 3), to eliminate the confounding opinions and concerns and only to give reliable evidence.

Acknowledgements I gratefully thank Dr Crapanzano who contributed in preparing some figures for this chapter.

References

Abraham E, Marincola FM, Chen ZN, Wang XD (2012) Clinical and translationalmedicine: integrative and practical science. Clin Trans Med 1:1

Aggarwal S, Pittenger MF (2005) Human mesenchymal stem cells modulate allogeneic immune cell responses. Blood 105:1815–1822

Asahara T, Murohara T, Sullivan A, Silver M, van der Zee R, Li T et al (1997) Isolation of putative progenitor endothelial cells for angiogenesis. Science 275:964–967

Atala A (2012) Tissue engineering of reproductive tissues and organs. Fertil Steril 98:21–29

Balistreri CR (2015) Genetic contribution in sporadic thoracic aortic aneurysm? Emerging evidence of genetic variants related to TLR-4-mediated signaling pathway as risk determinants. Vascul Pharmacol 74:1–10. doi:10.1016/j.vph.2015.09.006

Balistreri CR, Buffa S, Pisano C, Lio D, Ruvolo G, Mazzesi G (2015) Are endothelial progenitor cells the real solution for cardiovascular diseases? Focus on controversies and perspectives. BioMed Research International, Egypt Article ID 835934; 1–18

Balistreri CR, Madonna R, Melino G, Caruso C (2016) The emerging role of Notch pathway in ageing: focus on the related mechanisms in age-related diseases. Ageing Res Rev 29:50–65. doi:10.1016/j.arr.2016.06.004

Boyle AJ, Schulman SP, Hare JM, Oettgen P (2006) Is stem cell therapy ready for patients? Stem Cell Therapy for Cardiac Repair Ready for the Next Step. Circulation 114:339–352

Broxmeyer HE (2010) Cord blood hematopoietic stem cell transplantation. 2010 May 26. StemBook [Internet]. Harvard Stem Cell Institute, Cambridge; 2008. Available from http://www.ncbi.nlm.nih.gov/books/NBK44751/

Caliceti C, Nigro P, Rizzo P, Ferrari R (2014) ROS, Notch, and Wnt signaling pathways: crosstalk between three major regulators of cardiovascular biology. Biomed Res Int. 2014:318714. doi:10.1155/2014/318714

Chen SL, Fang WW, Ye F, Liu YH, Qian J, Shan SJ, Zhang JJ, Chunhua RZ, Liao LM, Lin S, Sun JP (2004) Effect on left ventricular function of intracoronary transplantation of autologous bone marrow mesenchymal stem cell in patients with acute myocardial infarction. Am J Cardiol 1(94):92–95

Duda DG, Cohen KS, Scadden DT, Jain RK (2007) A protocol for phenotypic detection and enumeration of circulating endothelial cells and circulating progenitor cells in human blood. Nat Protoc 2:805–810

Feyen DA, Gaetani R, Doevendans PA, Sluijter JP (2016) Stem cell-based therapy: Improving myocardial cell delivery. Advanced drug delivery reviews 106:104–115 pii:S0169-409X(16)30133–8

Gersh BJ, Simari RD, Behfar A, Terzic CM, Terzic A (2009) Cardiac cell repair therapy: a clinical perspective. Mayo Clin Proc 84:876–892. doi:10.1016/S0025-6196(11)60505-3

Ito H, Rovira II, Bloom ML, Takeda K, Ferrans VJ, Quyyumi AA, Finkel T (1999) Endothelial progenitor cells as putative targets for angiostatin. Cancer Res 59:5875–5877

Karcher JR, Hoffmann BR, Liu P, Liu Y, Liang M, Greene AS (2015) Genome-wide epigenetic and proteomic analysis reveals altered Notch signaling in EPC dysfunction. Physiological Rep 3(4):e12358. doi:10.14814/phy2.123 pii

Kwon SM, Alev C, Asahara T (2009) The role of notch signaling in endothelial progenitor cell biology. Trends Cardiovasc Med 19:170–173. doi:10.1016/j.tcm.2009.10.002

Li DW, Liu ZQ, Wei J, Liu Y, Hu LS (2012) Contribution of endothelial progenitor cells to neovascularization. Int J Mol Med 30:1000–1006. doi:10.3892/ijmm.2012.1108

Lindroos B, Suuronen R, Miettinen S (2011) The potential of adipose stem cells in regenerative medicine. Stem Cell Rev. 7:269–291. doi:10.1007/s12015-010-9193-7

Ma F, Morancho A, Montaner J, Rosell A (2015) Endothelial progenitor cells and revascularization following stroke. Brain research 1623:150–159. doi:10.1016/j.brainres.2015.02.010 pii: S0006-8993(15)00094-3

Mason C, Dunnill P (2008a) A brief definition of regenerative medicine. Regen Med. 3:1–5. doi:10.2217/17460751.3.1.1

Mason C, Dunnill P (2008b) The crucial linkage required between regenerative medicine bioprocessors and clinicians. Regen Med. 3:435–442. doi:10.2217/17460751.3.4.435

Medina RJ, O'Neill CL, Sweeney M, Guduric-Fuchs J, Gardiner TA, Simpson DA, Stitt AW (2010) Molecular analysis of endothelial progenitor cell (EPC) subtypes reveals two distinct cell populations with different identities. BMC Med Genomics 3:18. doi:10.1186/1755-8794-3-18

Murray IR, West CC, Hardy WR, James AW, Park TS, Nguyen A, Tawonsawatruk T, Lazzari L, Soo C, Péault B (2014) Natural history of mesenchymal stem cells, from vessel walls to culture vessels. Cell Mol Life Sci 71:1353–1374. doi:10.1007/s00018-013-1462-6

Nussbaum J, Minami E, Laflamme MA, Virag JA, Ware CB, Masino A, Muskheli V, Pabon L, Reinecke H, Murry CE (2007) Transplantation of undifferentiated murine embryonic stem cells in the heart: teratoma formation and immune response. FASEB J. 21:1345–1357

Okano H, Nakamura M, Yoshida K, Okada Y, Tsuji O, Nori S, Ikeda E, Yamanaka S, Miura K (2013) Steps toward safe cell therapy using induced pluripotent stem cells. Circ Res 112:523–533

Pelosi E, Castelli G, Testa U (2014) Endothelial progenitors. Blood Cells Mol Dis 52:186–194. doi:10.1016/j.bcmd.2013.11.004

Pittenger MF, Martin BJ (2004) Mesenchymal stem cells and their potential as cardiac therapeutics. Circ Res 95:9–20

Pompilio G, Cannata A, Peccatori F, Bertolini F, Nascimbene A, Capogrossi MC, Biglioli P (2004) Autologous peripheral blood stem cell transplantation for myocardial regeneration: a novel strategy for cell collection and surgical injection. Ann Thorac Surg 78:1808–1812

Porada CD, Atala AJ, Almeida-Porada G (2015) The hematopoietic system in the context of regenerative medicine. Methods 99:30057–30058. doi:10.1016/j.ymeth.2015.08.015 pii: S1046–2023

Povsic TJ, Najjar SS, Prather K, Zhou J, Adams SD, Zavodni KL, Kelly F, Melton LG, Hasselblad V, Heitner JF, Raman SV, Barsness GW, Patel MR, Kim RJ, Lakatta EG, Harrington RA, Rao SV (2013) EPC mobilization after erythropoietin treatment in acute ST-elevation myocardial infarction: the REVEAL EPC substudy. J Thromb Thrombolysis 36:375–383. doi:10.1007/s11239-013-0944-6

Qu K, Wang Z, Lin XL, Zhang K, He XL, Zhang H (2015) MicroRNAs: Key regulators of endothelial progenitor cell functions. Clin Chim Acta 448:65–73. doi:10.1016/j.cca.2015.06.017

Rajasekar P, O'Neill CL, Eeles L, Stitt AW, Medina RJ (2015) Epigenetic Changes in Endothelial Progenitors as a Possible Cellular Basis for Glycemic Memory in Diabetic Vascular Complications. J Diabetes Res. 2015:436879. doi:10.1155/2015/436879

Resch T, Pircher A, Kähler CM, Pratschke J, Hilbe W (2012) Endothelial progenitor cells: current issues on characterization and challenging clinical applications. Stem Cell Rev. 8:926–939. doi:10.1007/s12015-011-9332-9

Rouchi AH, Mahdavi-Mazdeh M (2015) Regenerative medicine in organ and tissue transplantation: shortly and practically achievable? Int J Organ Trans Med 6(3):93–98

Sherman W, Martens TP, Viles-Gonzalez JF, Siminiak T (2006) Catheter-based delivery of cells to the heart. Nat Clin Pract Cardiovasc Med. 3(Suppl 1):S57–S64

Sieveking DP, Buckle A, Celermajer DS, Ng MK (2008) Strikingly different angiogenic properties of endothelial progenitor cell subpopulations: insights from a novel human angiogenesis assay. J Am Coll Cardiol 51(6):660–668. doi:10.1016/j.jacc.2007.09.059

Stoltz JF, de Isla N, Li YP, Bensoussan D, Zhang L, Huselstein C et al (2015) Stem cells and regenerative medicine: myth or reality of the 21th century. Stem Cells Int. 2015:734731. doi:10.1155/2015/734731

Thomas ED, Epstein RB (1965) Bone marrow transplantation in acute leukemia. Cancer Res 25:1521–1524

Urbich C, Dimmeler S (2004) Endothelial progenitor cells functional characterization. Trends Cardiovasc Med 14:318–322

Van Craenenbroeck EM, Van Craenenbroeck AH, van Ierssel S, BruyndonckxL Hoymans VY, Vrints CJ et al (2013) Quantification of circulatingCD34+/KDR+/CD45dim endothelial progenitor cells: analytical considerations. Int J Cardiol 167:1688–1695. doi:10.1016/j.ijcard.2012.10.047

van Ierssel SH, Jorens PG, Van Craenenbroeck EM, Conraads VM (2014) The endothelium, a protagonist in the pathophysiology of critical illness: focus on cellular markers. Biomed Res Int. 2014:985813. doi:10.1155/2014/985813

Wang X (2012) A new vision of definition, commentary, and understanding in clinical and translational medicine. ClinTransl Med 1:5. doi:10.1186/2001-1326-1-5

Watt SM, Gullo F, van der Garde M, Markeson D, Camicia R, Khoo CP, Zwaginga JJ (2013) The angiogenic properties of mesenchymal stem/stromal cells and their therapeutic potential. Br Med Bull 108:25–53. doi:10.1093/bmb/ldt031

Wojakowski W, Landmesser U, Bachowski R, Jadczyk T, Tendera M (2012) Mobilization of stem and progenitor cells in cardiovascular diseases. Leukemia 26:23–33. doi:10.1038/leu.2011.184

Chapter 2
Endothelial Progenitor Cells and Their Clinical Applications as Potential Disease Biomarkers and Therapeutic Agents: Evidence and Controversies Regarding Their Effectiveness

Abstract The discovery and progress in the field of endothelial progenitor cells (EPCs) are leading numerous researchers to discuss applying them as potential predictive, diagnostic and prognostic biomarkers of age-related diseases in order to improve the prevention, management and outcome of such diseases. However, growing evidence suggests changes in the number and functions of EPCs significantly associated with cardiovascular diseases, diabetes, Alzheimer's disease, cancer and the associated complications. In addition, EPCs have been suggested as therapeutic agents for many of these diseases. Certainly, many aspects remain to be clarified, and controversies must be verified and eliminated. As a result, the road is still hard and long, but the advances obtained may offer us greater and unique opportunities to meet the considerable challenge of age-related diseases.

Keywords EPCs · Clinical applications · Disease biomarkers · Therapeutic agents · Age-related diseases

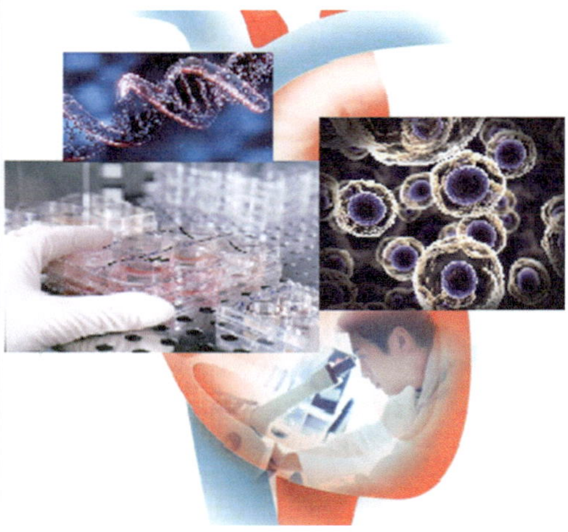

© Springer International Publishing AG 2017
C.R. Balistreri, *Endothelial Progenitor Cells*, UNIPA Springer Series,
DOI 10.1007/978-3-319-55107-4_2

The figure to the left shows *Thomas Sydenham, MD, English Physician, 1624–1689*. *He formulated the quote, "A man is as old as his arteries"*, demonstrating that physicians and scientists have focused their attention on the cardiovascular system for centuries; this interest had already arisen at the time of Dr *Thomas Sydenham. Since then, the* physicians and scientists *have underlined that the age of a man is not the chronological age, but rather the biological age, principally determined by the biological status of cardiovascular system. This has led the scientific community to study cardiovascular biology and vascular aging and to center its efforts on related mechanisms (as shown to the right of the figure).*

In particular, in recent years, studies have focused on "*endothelium dysfunction*" and the related cellular and molecular mechanisms. This is in accordance with well-recognized evidence that endothelium dysfunction is the first pathological condition associated not only with all cardiovascular diseases, but also with other age-related diseases, being the endothelial cells (ECs) components of the stroma of all tissues and organs. Accumulation of EC damages followed by death and impaired cardiovascular self-repair, due principally to altered age-related function of tissue-specific resident and circulating stem or progenitor cells, characterizes the onset of endothelial dysfunction. To be specific, when an imbalance between damage/death and regeneration of the endothelium occurs, the integrity and function of endothelium is threatened, and the origin and progression of CVDs, or other age-related diseases, is favored. Experimental studies on the endothelium have led to relevant and promising discoveries and advances. Among these findings, strong opinion has been emerging that the solution to counteract the onset of age-related diseases, has its roots in the complete knowledge of the biology of *stem and progenitor cells* (i.e. hematopoietic and non-hematopoietic stem cells and their progenitors), the related cellular mechanisms and the evolutionarily conserved signaling pathways of these cells, involved in the genesis of vessels ("*vasculogenesis*"), tissues, organs and systems (*as described and discussed in Balistreri et al., Ageing Res Rev. 2016;29:50–65. doi: 10.1016/j.arr.2016.06.004, see references*)

—Carmela Rita Balistreri.

2.1 Endothelial Dysfunction as the Early Event of Chronic Age-Related Diseases: EPCs as Cellular Biomarkers

The endothelium has critical roles as a barrier and primary sensor of physical and chemical changes in the bloodstream. Essential for controlling the passage of fluid into tissue, the endothelium is also a determinant of cellular trafficking, an inhibitor of coagulation, and a regulator of blood pressure through the release of vasodilatory and constricting agents such as nitric oxide (NO) and endothelin. Endothelial dysfunction is an encompassing term for a shift from a normal, healthy endothelium to a stressed/damaged endothelium with a pro-vasoconstriction, pro-coagulation and pro-inflammatory phenotype. It has been documented for many years that endothelial dysfunction is the first pathological condition for all cardiovascular diseases (CVDs). Today, endothelial dysfunction has been also observed and associated with other age-related diseases, being the endothelial cells (ECs) components of the stroma of all tissues and organs. Accumulation of EC damages followed by death and impaired cardiovascular self-repair, due principally to altered age-related function of tissue-specific resident and circulating stem or progenitor cells, characterizes the onset of endothelial dysfunction (see Fig. 2.1) (Cahill and Redmond 2016; Regina et al. 2016). To be specific, when an imbalance between damage/death and regeneration of the endothelium occurs, integrity and function of endothelium is threatened, and the origin and progression of CVDs, or other age-related diseases, is favored (Madonna et al. 2016).

Endothelial dysfunction in humans can be assessed by invasive and non-invasive techniques, the most popular being flow-mediated vasodilation. However, these functional methods provide no information on the mechanism of dysfunction, which may be useful in guiding therapy. Biochemical and cellular surrogates of endothelial dysfunction may provide some insights into the health status of the endothelium. In particular, EPCs, may provide information about endothelial cell homeostasis and mechanisms underlying endothelial dysfunction, but they may also represent potential predictor, diagnostic and prognostic biomarkers of age-related diseases in order to improve their prevention, management and outcome. An increase in the number of circulating endothelial cells (CECs) and EPCs in the peripheral blood (PB) is normally observed in cases of vascular injury, repair and neovascularization. These cells rarely exist in the PB of healthy individuals (as reported in Chap. 1). Therefore, when they are present in the PB of individuals, their phenotypes and quantity in the PB may serve as surrogate diagnostic or prognostic parameters of vascular injury, repair and/or as an indication of tumor growth. Indeed, higher circulating levels of progenitor cells reflect greater repair capacity, and have been shown to reduce the progression of atherosclerosis and ischemic vascular injuries, but co-temporally to indicate angiogenesis and tumor progression (Critser and Yoder 2010; Cahill and Redmond 2016; Madonna et al. 2016; Moccia et al. 2015; Regina et al. 2016). This last condition leads to the consideration of vasculogenesis mediated by EPCs as an intriguing therapeutic agent, both in the early and late stages of cancer progression (Gao et al. 2009; Moccia et al. 2015). Furthermore, it has been demonstrated that age, a significant risk factor for the development of age-related diseases, interestingly affects EPCs. EPCs are subject to age-associated changes that diminish their number in circulation and function (Williamson et al. 2012). Decreased EPC numbers have been also associated with metabolic syndrome, which confers a high risk for CVDs (Devaraj and Jialal 2012).

Given the critical role of the endothelium in vascular health and the importance of maintaining a ‘healthy’ endothelium as opposed to a ‘dysfunctional’ one, clinical benefit may be derived from harnessing EPCs that can differentiate into mature endothelial cells to use as therapy for vascular disease. With this in mind, strategies that have been employed to enhance EPC number and functionality include statins, ACE inhibitors, angiotensin receptor blockers and peroxisome proliferator-

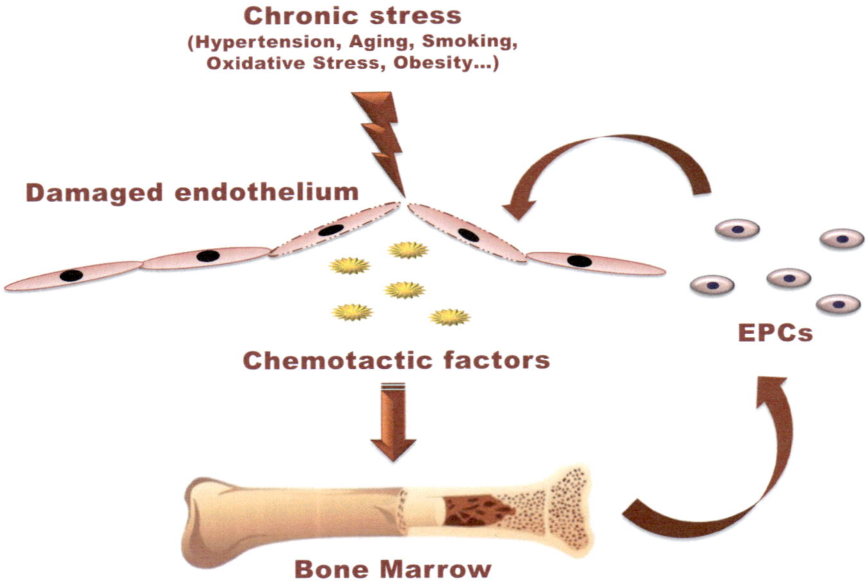

Fig. 2.1 Endothelium damage and dysfunction, and recruitment of EPCs. Several factors (aging, hypertension, oxidative stress, hyper-lipemia, obesity etc.) act as triggers of endothelium damage and induce endothelium dysfunction. This determines the release of chemiotactic factors that induce the recruitment of BM-EPCs in the sites of injury, restoring endothelium integrity and allowing normal function. However, different endogenous factors affect this capacity (as described in Fig. 2.2.)

activated receptor gamma (PPAR-g) agonists, although the mechanisms involved remain unclear (Devaraj and Jialal 2012). Other clinical approaches involve providing exogenous EPCs to the circulation to replenish numbers that are reduced in diseased patients (Fadini et al. 2010). Finally, the discovery of 'resident' stem/progenitor cells in the vessel wall, which can differentiate into all types of vascular cells (Tang et al. 2012), and the evidence that endothelial cells may transform into smooth muscle-like cells (Lao et al. 2015), have complicated the vascular 'progenitor cell' research field even further.

In this chapter, a larger number of these aspects are described and discussed.

2.2 Relationship Between Circulating EPC Levels and Age-Related Diseases

Consistent evidence shows that the defects in the mobilization, recruitment and function of EPCs occur in CVDs, diabetes, AD, cancer and other age-related pathologies (as previously mentioned). Changes in EPCs also have a pathogenic

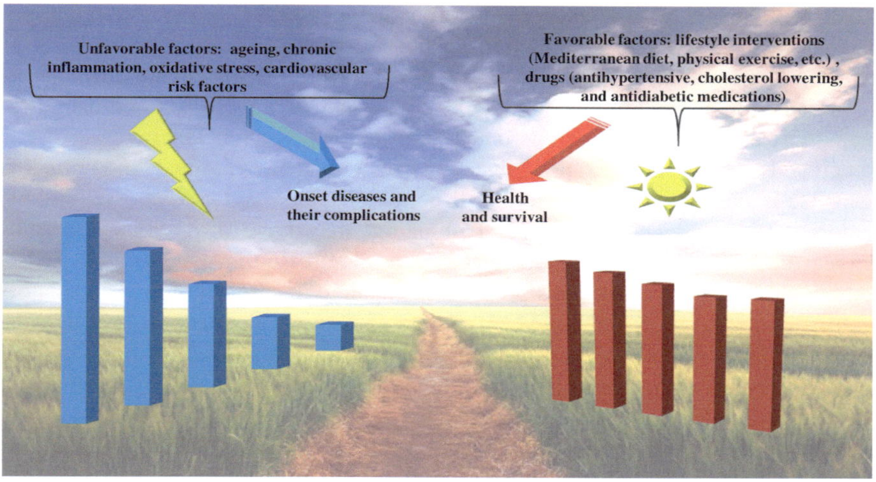

Fig. 2.2 Unfavorable and favorable factors modulating circulating EPC levels. Several factors modulate the circulating levels of EPCs. Physiological factors (e.g., gender and pregnancy) are associated with high levels of EPCs, as well as lifestyle interventions (e.g., Mediterranean diet, physical exercise and no smoking.) and drugs (e.g., statins). These factors might preserve the health of individuals and, consequently, allow them to achieve a very old age. By contrast, unfavorable factors, including the aging process and the related chronic inflammatory response associated with high oxidative stress, as well as the typical age-related risk factors (e.g., smoking, obesity, no physical exercise, diabetes and hyperlipidemia) trigger the number and function of EPCs. As a result, reduced levels of EPCs are observed in these conditions, which are significantly associated with onset of CVDs, diabetes and AD

role in the development of complications related to the diseases previously mentioned (Critser and Yoder 2010; Cahill and Redmond 2016; Madonna et al. 2016; Moccia et al. 2015; Regina et al. 2016). This relationship is confirmed by data from several current studies, which identified several factors able to modulate both the circulating levels and the functions of EPCs, consequently acting as unfavorable and favorable endogenous factors. In addition, exogenous interventions, such as drug therapies, nutrition interventions and lifestyle modifications, have been proposed as potential measures to improve circulating EPC levels and functions. For a clear knowledge on the potential relevance of EPCs in clinical area, it describes them and their effects on EPC number and function (Fig. 2.2).

2.2.1 Unfavorable Factors Associated with Decreased Circulating EPC Levels

EPC levels and functions have been shown to be significantly modulated by several factors (Fig. 2.2). Firstly, it has been demonstrated that aging (as previously mentioned) negatively influences EPC function and viability, by determining a decreased potentiality of endothelial repair (Felice et al. 2012; Williamson et al. 2012). The causes of this age-related condition have been principally attributed to age-related inflammation and oxidative stress, which appear to modulate EPC bioactivity and determine dysfunction (Lin et al. 2013; Rodiño-Janeiro et al. 2013). Specifically, EPC mobilization in cases of transient restricted inflammatory response is observed as a physiological condition. Persistent or excessive inflammatory responses, which are typical of aging, may have deleterious effects by decreasing EPC circulating numbers (Lin et al. 2013; Rodiño-Janeiro et al. 2013). Functional EPC activity is also significantly impaired in cases of high inflammatory stimulation, as in heart failure. Mechanisms regulating this effect are still unclear. However, convincing evidence suggests that prolonged exposure of bone marrow (BM) to increased pro-inflammatory stimulation (chronic response typical of aging) may determine EPC pool exhaustion. In this condition, a small number of EPCs, prevalently immature or dysfunctional, might be released. However, existing clinical evidence regarding the association of inflammation with reduced EPC levels is largely circumstantial and observational (Lin et al. 2013; Rodiño-Janeiro et al. 2013).

As mentioned in the Chap. 1, oxidative stress may also play a crucial role in EPC mobilization from BM and functional bioactivity (see Fig. 1.9). Reactive oxygen species (ROS) exert a direct cytotoxic effect on the vascular endothelium. Increased superoxide generation reduces EPC levels and impairs EPC function, as demonstrated by increased apoptosis and reduced EPC number after incubating with high levels of hydrogen peroxide (H_2O_2) (Lin et al. 2013; Rodiño-Janeiro et al. 2013).

An increasing body of evidence also suggests that cardiovascular risk factors (smoking, diabetes, hypertension, lipid disorders, abdominal obesity, metabolic syndrome, etc.) affect EPC number and proprieties (Werner and Nickenig 2006) (Fig. 2.2).

Endocrine disorders, such as hyperparathyroidism and hypothyroidism, may also alter EPC levels (as summarized in Balistreri et al. 2015).

2.2.2 *Physiological Factors and High Circulating EPC Levels*

Physiological factors also influence EPC circulating levels and their function. Firstly, gender appears to modulate EPC levels, as demonstrated by Fadini and colleagues (Fadini et al. 2008). Women have higher EPC levels than men and estrogen acts as the physiological stimuli determining these useful effects (Fadini et al. 2008) (Fig. 2.2). Another significant factor is pregnancy, which characterized by high circulating levels of EPC (Sugawara et al. 2005).

2.2.3 *Exogenous Measures as Potential Strategies for Increasing Circulating EPC Levels*

Drug therapies can significantly influence EPC levels and function, acting as anti-inflammatory and antioxidant factors. Recently, a research group has found evidence that there is a significant interaction between cardiovascular pharmacotherapies and the improvement of EPC number and function. In particular, they described the effects observed in clinical studies on EPC number and function from patients with different CVDs and treated with different medications, including antihypertensive, cholesterol lowering, and antidiabetic medications (Lee and Poh 2014) (Fig. 2.2).

Nutrition interventions and lifestyle modifications may determine increases in EPC levels and potentiate their function (Fig. 2.2). In particular, the Mediterranean diet seems to act in a beneficial manner on circulating EPC levels and function (Fernández et al. 2012). Similarly, physical exercise seems to lead to an improvement in circulating EPC levels (see Fig. 2.2) (Williamson et al. 2012). This has been found in both healthy subjects and patients affected by CVDs. Thus, even in patients, with diffuse atherosclerosis and multiple risk factors, reparative capacity dependent on circulating BM-derived EPC is retained and can be enhanced in a physiological way (Williamson et al. 2012).

2.3 Correlation Between Blood EPC Levels and Cardiovascular Profile

As mentioned, a strict relationship exists between EPC circulating levels and cardiovascular risk profile. This may be of clinical relevance, and possible new recommendations and preventive CVD measures may arise as a result. Indeed, in 2003, Hill and colleagues reported in a notorious journal that the circulating levels of EPCs represent a better predictor of vascular reactivity than the conventional cardiovascular risk (Hill et al. 2003). In addition, they found that in vitro EPC senescence reflects the risk profile of the donor's subjects. Thus, the authors suggested that EPCs might be considered to be an optimal biomarker for vascular function and cardiovascular risk (Hill et al. 2003). Abnormalities in circulating EPC levels and function have been observed in a large number of studies on different CVDs (Fig. 2.3). As result, EPCs might be suggested to be predictive, diagnostic and prognostic CVD biomarkers (Favero et al. 2014). In order to substantiate this role of EPCs, several pathological cardiovascular conditions are discussed in the following paragraphs.

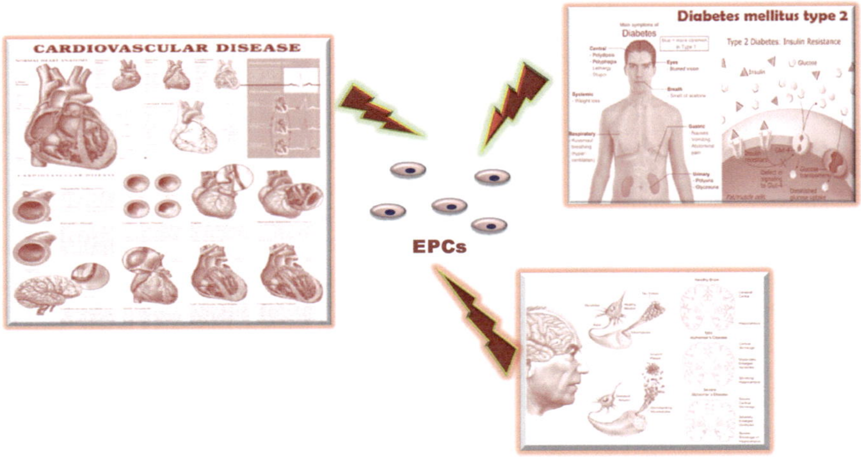

Fig. 2.3 Alterations in reduction of both EPC number and function in CVDs, diabetes and Alzheimer's disease

2.3.1 *Atherosclerosis*

Circulating EPC numbers are a surrogate marker for cumulative cardiovascular risk (Sen et al. 2011). In fact, low circulating EPC numbers related to hyperlipidemia and vascular inflammation are an independent predictor for endothelial dysfunction and atherosclerosis. Schmidt-Lucke et al. (2005) demonstrated that reduced levels of circulating EPCs independently predict atherosclerotic disease progression, supporting an important role for endogenous vascular repair to modulate the clinical course of coronary artery disease (CAD). Reciprocally, correction of hyper-lipidemia in cardiovascular patients correlates with an increase in circulating EPC levels and improved cardiovascular risk profile (Croce et al. 2006; Vasa et al. 2001). On the other hand, hypercholesterolemia is the major event initiating atherosclerosis. It induces disruption of vascular endothelium integrity.

In settings of endothelial damage, EPCs are mobilized from BM into circulation and are home to sites of vascular injury, where they aid endothelial regeneration and, subsequently, regression of atherosclerosis.

2.3.2 *Hypertension*

Hypertension remains one of the major risk factors for CVD and cerebrovascular diseases. EPCs are involved in the pathophysiology of hypertension and in cardiac fibrosis in adult ischemic injury and hypertensive heart disease. Eirin et al. (2013) showed that, despite preserved kidney function and controlled blood pressure, renal vein and circulating inflammatory markers were elevated in black essential hypertensive (EH) patients and correlated with increased inflammatory endothelial cells (IECs) and decreased EPC levels. Increased release of cytokines and IECs in black EH patients may impair EPC reparative capacity and create a predisposition to hypertensive vascular injury. This process may aggravate vascular damage and accelerate hypertension-related morbidity and mortality rates in black EH patients. At the same time, according to a recent study performed by Braitsch et al. (2013), a particular phenotype of EPC, Tcf 21 positive, has been involved in cardiac fibrosis associated with hypertension.

2.3.3 Coronary Artery Disease (CAD)

A large number of studies analyzed the prognostic effects of EPCs in the context of CAD. Accordingly, Padfield et al. (2013) suggested that traditional EPC populations were not related to the extent of CAD or the clinical outcome. On the contrary, Flammer et al. (2012) highlighted a strong correlation between circulating OCN+ 'early' EPCs and unstable atherosclerosis. Furthermore, D'Amario et al. (2013) assessed that the function of resident EPCs provides critical information concerning the recovery of the myocardium following successful revascularization of patients who have chronic CAD. Negative left ventricular (LV) remodeling over a period of 12 months after coronary bypass was not observed if the EPCs' compartment before surgery retained a significant growth reserve. A decline in the replicative potential of EPCs was paralleled by alterations in ventricular wall thickening, together with chamber dilation and reduction in LV mass-to-chamber volume ratio.

2.3.4 Bicuspid Aortic Valve, Aortic Valve Regurgitation (AR) and Aneurysm

Vaturi et al. (2011) recently showed that patients with bicuspid aortic valve and significant valvular dysfunction exhibit circulating EPCs with impaired functional properties. At the same time, Shimoni et al. (2013) concluded that there are decreased numbers of circulating EPCs and late apoptotic EPCs in patients with AR compared with a control group of subjects with similar risk factors and no valve disease. This observation was evident in patients with different AR aetiologies. In patients with AR, EPC number correlated negatively with LV thickness, and late apoptotic EPCs correlated negatively with male sex and LV end systolic diameter. In addition, EPCs seems to have an important role in aneurysm physiopathology and in neo-intima formation and re-endothelialization in rabbit elastase-induced aneurysm after flow diverter treatment (Van Spyk et al. 2013).

2.3.5 *Cerebral Ischemia*

In stroke patients, EPC numbers are decreased and inversely correlate with age, stroke type and functional stroke outcome. Taguchi et al. (2004) showed that diminished numbers of circulating immature cells might impair such physiological and reparative processes, potentially contributing to cerebrovascular dysfunction. In particular, there is a strong inverse correlation between numbers of circulating CD34- and CD133-positive cells and cerebrovascular infarctions. On the other hand, Bogoslovsky (2010) revealed that low EPC numbers indicate large lesion volumes and higher EPC levels are linked with small ischemic lesions and smaller lesion growth. Despite the presence of various risk factors, comorbidity and medications, EPC levels remained predictive of lesion and lesion growth volume.

2.3.6 *Other CVD Diseases*

The role of EPCs have been investigated in other cardiac pathologies and peripheral vascular disease. Chan et al. (2013) discovered that in patients with PR interval elongation syndrome, circulating CD133(+)/KDR(+) EPC level was significantly increased. Oterino et al. (2013) found that CD62E+ EPCs might be considered to be a marker for vascular damage in migraine patients. Condorelli et al. (2013) have analyzed the role of EPC in arterial erectile dysfunction: low pharmacological response to phospho-diesterase type 5 inhibitors is associated with a significantly higher level of endothelial apoptosis, which is associated with higher serum concentrations of circulating late immunophenotype of EPCs. Rodrigo et al. (2013) assessed that low levels of EPCs in patients with sickle cell disease increased the risk of cardiovascular disease.

2.4 EPCs in Alzheimer Disease

Vascular endothelial dysfunction plays a key role in the pathogenesis of Alzheimer's disease (AD). Patients with AD have displayed decreased circulating EPCs, which repair and maintain the endothelial function (Hayashi et al. 2009; Lee et al. 2009; Kong et al. 2011). Transplantation of EPCs has emerged as a promising approach for the management of cerebrovascular diseases including ischemic stroke; however, its impact on AD has been poorly described (Safar et al. 2016) (Fig. 2.3).

2.5 Diabetes and EPCs

EPC alterations have been observed in Type 2 diabetes mellitus (T2DM) (Fig. 2.3). Fadini and colleagues first proved that diabetic animals showed defective EPC mobilization and compensatory angiogenesis in a hind limb ischemia-reperfusion model (Fadini et al. 2013). In addition, circulating EPCs levels were found to be negatively correlated with the progression of vascular-associated complications of diabetes, including diabetic retinopathy (Ling et al. 2012), nephropathy and wound healing (Caballero-Villarraso 2012, Caballero et al. 2013). T2DM is proved to be related to decreased levels and impaired function of circulating EPCs, including decreased proliferation, migration abilities and increased apoptosis, resulting in a reduced capacity for angiogenesis and tissue recovery (Ling et al. 2012). Alba et al. (2013) noted differential relations between EPCs and outcomes in patients with versus those without diabetes. Higher EO-CFUs (early-outgrowth colony-forming units) and lower CD34+ VEGFR2+ cells were associated with improved functional capacity and reduced mortality in non-diabetic patients. In patients with diabetes, lower EO-CFUs were associated with improved functional capacity.

In identifying the hypothetical causes related to reduced levels and functions of EPCs in diabetes, it is emerging that disorders of glucose regulation are associated with abnormalities in EPC biology, including reduced circulating numbers of EPCs, defective mobilization from BM and impaired functional properties of EPCs. In particular, several of the abnormalities associated with insulin resistance, including reduced nitric oxide (NO) bioavailability, increased production of ROS (reactive oxygen species) and down-regulation of intracellular signaling pathways, have the potential to disrupt EPC function. However, these observations merit additional investigations (Cubbon et al. 2009; Altabas 2015).

2.6 EPCs in Solid and Hematological Cancers, and Recent Controversies Relating to Their Incorporation into Tumor Neovessels

As described previously, changes in EPCs functions and levels are associated with both onset and complications of CVDs, diabetes and AD. Thus, restoring EPC functions in these diseases may prevent disease development and complications. By contrast, targeting EPCs in cancer might also have potential therapeutic benefits. In this regard, it is important to clarify that EPCs actively contribute to tumor-angiogenesis, as demonstrated for the first time by Lyden et al. (2001). This finding was confirmed by Peters et al. (2005). They determined the role of such cells in human tumor angiogenesis, by studying six individuals who developed cancers after BM transplantation with donor cells derived from individuals of the opposite sex. By performing fluorescence in situ hybridization (FISH) with sex chromosome-specific probes in conjunction with fluorescent antibody staining, they found that EPCs contributed to tumor endothelium, but at low levels, averaging only 4.9% of the total. More recently, Nolan and coworkers used BM-progenitor cells isolated from GFP+ mice and injected them into lethally irradiated syngenic wild-type recipients to investigate the contribution of EPCs to tumor angiogenesis (Nolan et al. 2007). Reconstituted animals were xenografted with three distinct tumor types—namely B6RV2 lymphoma, Lewis lung carcinoma (LLC) and melanoma—and then examined at various stages of tumor development with the use of endothelial (VE cadherin, CD31, endoglin and VCAM), hematopoietic (CD11b, CD45RB, CD41) and progenitor (CD133) markers. GFP expression, in turn, ensured BM origin of vessel cells. These authors first found that BM-derived GFP + cells were recruited at the periphery of LLC at the early stages of tumor growth (days 4–6) prior to the sprouting of endothelial cells from nearby capillaries. These cells were identified as EPCs based on their morphological and phenotypic characterization. When LLC tumors were inspected at later stages (6–8 days), they showed chimeric vessels comprising both non-BM derived cells and BM-derived GFP+ EPCs. Importantly, high-resolution stereo-confocal microscopy confirmed that GFP+ cells did not occupy a perivascular location, while optical sectioning of multiple z-stacks (30 μm resolution) displayed that BM-derived endothelial cells possess a single nucleus and that CD31 (indicative of endothelial origin) and GFP signals derive from the same individual cell. This proved that tumor endothelial cells could actually originate from BM-mobilized EPCs. Intriguingly, flow cytometric analysis revealed that the percentage of BM-derived EPCs (GFP+ VE-cadherin+ CD31low CD11b−) decreases from 25–35% in the early phase of tumor development (4–6 days) to 6–8% at later stages (6–8 days), while the fraction of local non-BM-derived endothelial cells (GFP− VE-cadherin+ CD31low CD11b−) increased to 65–75% at days 10–14 (Nolan et al. 2007). The same results were found in a transgenic breast cancer mouse model (MMTV-PyMT) (Nolan et al. 2007). Thus, EPCs play a crucial role during the initial steps of tumor vascularization. The MMTV-PyMT transgenic mice were further exploited to assess EPC contribution to the dynamics of the vessel assembly that turns dormant micrometastases into lethal macrometastases (Gao et al. 2008). By using the same procedure described in their seminal paper (Nolan et al. 2007), Gao and coworkers focused on the angiogenic switch in lung metastases that spontaneously develop in this breast cancer model. They found that micrometastases formed by week 12 and were poorly vascularized, as shown by the lack of CD31+ vessels. Nevertheless, macrometastases that appeared at week 16 were positive to CD31 staining and displayed luminally incorporated BM-derived GFP+ endothelial cells in about 11% of neovessels (Nolan et al. 2007; Gao et al. 2009). This means that EPCs are home to micrometastatic foci and contribute to neovessel formation, thereby sustaining the macrometastatic transition. The low percentage of EPC engraftment suggested that, apart from a structural role, they drive the angiogenic switch in a paracrine manner. The same findings were obtained by analyzing lung metastases in LLC xenograft mice. Again, by using this model, the authors further found that many BM-derived GFP+ cells are recruited to micrometastases, but confocal microscopy analysis revealed that only endothelial cells (GFP+ CD31+) integrated into neovessels. More specifically, BM-derived EPCs were recruited at the outer rim of the metastatic lesion, while hematopoietic stem cells (HSCs) adopted a perivascular location (Nolan et al. 2007). Consistent with these observations, the acute and conditional short hairpin RNA (shRNA)-mediated genetic ablation of Id1 in BM-derived EPCs did not decrease the number of micrometastatic lesions but prevented the macrometastatic transformation in LLC xenografts (Gao et al. 2008).

The studies, above reported, emphasize a strong relation between EPC cells and angiogenesis during tumor progression studies emphasize a strong relationship between EPCs and in tumorogenesis (Critser and Yoder 2010). Accordingly, EPC recruitment during tumor progression significantly contributes to the cancer-associated neovasculature. Thus, targeting these cells could be effectively used to halt tumor progression by preventing vascularization of neoplastic deposits. In order to augment understanding of this role, efforts have been focused on identifing the related- mechanisms. In terms of this, it is helpful to consider that EPC-mediated vasculogenesis is a multistep process, which includes: (a) mobilization of EPCs from BM; (b) active arrest and transendothelial extravasation of EPCs into the interstitial space of the growing tumor; and (c) EPC incorporation of into neovessels or paracrine support of the nascent microvasculature. In the context of cancer, tumor-derived signals instigate the BM compartment to mobilize EPCs and recruit them to the tumor bed (Fig. 2.4). For example, the high levels of VEGF produced by tumors mobilize BM-resident EPCs to enter the peripheral circulation, and enhance the recruitment of these cells to the tumor sites (Fig. 2.4). VEGF mobilizes these cells by interacting with the VEGFR-1 and VEGFR-2 receptors

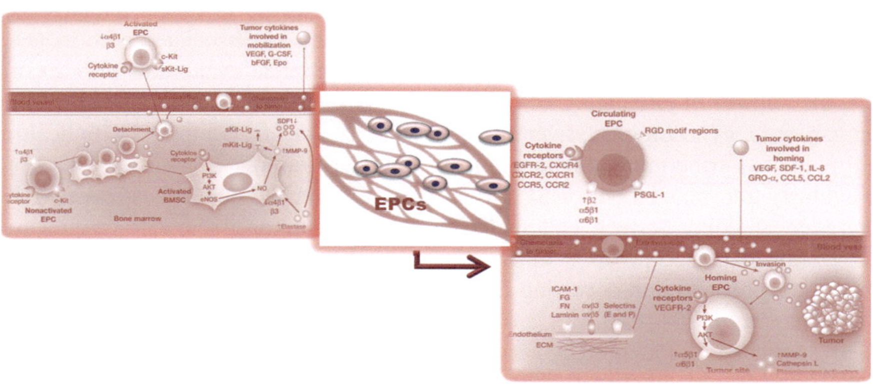

Fig. 2.4 Recruitment of EPCs in neoplastic sites. In cancer, tumor-derived signals instigate the BM compartment to mobilize EPCs and recruit them to the tumor bed. Thus, EPCs contribute to neovessel formation, *angiogenesis process* and cancer progression with metastasis formation

(R) that are expressed on EPCs. VEGF also induces MMP-9 expression by tumor cells, and results in the release of stem cell active soluble kit ligand, which in turn forces the translocation of quiescent VEGFR2+ c-kit+ EPCs into the circulation for recruitment to sites of ischemia, inflammation or tumor growth (de la Puente et al. 2013a, b; Moschetta et al. 2014) (Fig. 2.4).

Hypoxia also mobilizes EPCs: malignant tumor growth results in neoplastic tissue hypoxia, which then mobilizes EPCs in a paracrine fashion. EPCs widely express CXCR4, which is the receptor for stromal cell-derived factor-1α (SDF-1α) and a member of the chemokine CXC subfamily. Hypoxia induces SDF-1α secretion by HIF in tumor cells; this mobilizes EPCs in the BM. To date, both SDF-1α/CXCR4 and VEGF/VEGFRs pathways are the main known mediators of EPC BM mobilization during cancer development, and represent potential targets for new anti-vasculogenic therapies (de la Puente et al. 2013a, b; Moschetta et al. 2014).

The contribution of EPCs to neovessel formation has been reported for several spontaneous, transplanted and metastatic solid tumor models; however, other studies that use some of these same tumor models, report almost no incorporation of BM-derived EPCs to the nascent vasculature, and point instead to the possible provasculogenic role of other BM-derived cell populations of hematopoietic origin. It is also possible that incorporation of EPCs into the tumor vasculature is tumor-stage specific, and that the relative contribution of these cells to tumor-associated neovessels is greater at earlier rather than later stages of tumor growth. In growing tumors, incorporated EPCs could eventually be replaced with non-BM-derived endothelial cells, after the angiogenic process has taken place. In addition, EPC-induced neovessel formation depends on the tumor type and its related stage specificity. Thus, the selection of specific mouse cancer models may be important for analyzing EPC-mediated neovascularization, which may take place at earlier or later stages, depending on the specific tumor models (de la Puente et al. 2013a, b; Moschetta et al. 2014).

In the clinical setting, higher levels of circulating EPCs in cancer patients correlate with increased angiogenesis and metastases, and are associated with reduced patient survival, highlighting the importance of preclinical evaluations. For example, the number of circulating EPCs is reportedly increased in patients with breast, lung, gastric, gynecological, liver and glioblastoma cancers, compared with healthy individuals. Furthermore, examination of secondary tumors in patients who were previously transplanted with HSCs from a sex-mismatched donor, has revealed the presence of donor-derived endothelial cells in the tumor-associated vasculature, suggesting that incorporation of BM-derived EPC into vessels maybe a relevant biological phenomenon occurring in a cancer patient (de la Puente et al. 2013a, b; Moschetta et al. 2014).

In summary, although many issues remain to be clarified, preclinical and clinical data consistently highlight the importance of EPCs (including subsets of BM-derived cells of hematopoietic origin) in mediating the progression of solid tumors, including metastatic dissemination.

The full impact of vasculogenesis in the pathogenesis of hematological cancers has yet to be clarified, primarily due to technical reasons including the lack of a consensus definition of EPCs themselves and lack of standardized methodologies for the quantification of these cells. Indeed, recent studies have established that most of the cells previously identified as circulating EPCs likely represent provasculogenic hematopoietic cells at various stages of differentiation; this may account for many of the biological activities actually attributed to putative EPCs (Moschetta et al. 2014).

In addition, several research groups recently questioned EPC participation in the angiogenic switch, even if there is undoubted evidence in favor of EPC involvement in tumor growth and metastatization, as previously described. On the other hand, the initial enthusiasm raised by preclinical investigations on established murine models of different human cancers has not been followed by a successful translation into oncological practice (see below in Sect. 2.8).

2.7 EPCs as Therapeutic Agents

Among BM stem and progenitor cells, EPCs actually are the object of study for cell therapy in the treatment of several age-related diseases (Fig. 2.5). Interesting data have been and are being obtained. However, there has previously been a lack of a consistent evidence in this arena, which still hampers the clinical application of these cells, as extensively seen in several recent studies. The next sections describe the application of EPCs as therapeutic agents in age-related diseases.

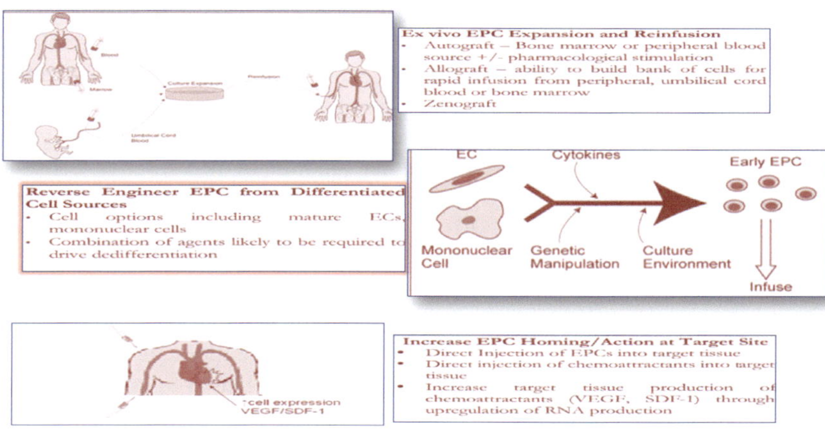

Fig. 2.5 Endothelial progenitor cells as cell therapy and different methods of injection

2.7.1 EPCs as Therapeutic Agents for CVDs: Focus on Ischemic Disease Treatment, Capacity and Doubts

As it is well known, the treatment of acute myocardial infarction (AMI) and ischemic cardiomyopathy includes rapid revascularization to limit ischemic damage and consecutive LV dysfunction and remodeling, and optimized secondary prevention strategies aiming to attenuate progression of cardiac dysfunction and vascular disease. From this point of view, some current procedures, such as percutaneous coronary intervention (PCI) and coronary artery bypass grafting (CABG) seem to excel than the unresponsive pharmacological therapy, by making the combination of prevention and therapeutic procedures a complementary union of partners in a continuum of management. Despite these therapeutic approaches, the prevalence of heart failure from post ischemic cardiac dysfunction has increased, and is associated with a substantial morbidity. Indeed, there is a substantial number of patients with ischemic heart disease that is refractory to current therapeutic approaches, and who are lacking further treatment options, i.e. patients with angina pectoris and no option of interventional or surgical revascularization. This condition commonly occurs in old patients. With aging, there is an increase in the incidence and severity of ischemic cardiovascular diseases. Pharmacotherapeutic regimens as well as revascularization therapy, such as CABG, are not sufficient to bring about an improvement of a widely impaired cardiac function. Thus, in the absence of effective endogenous repair mechanisms after cardiac injury, cell-based therapies have rapidly emerged as a potential novel therapeutic approach in ischemic heart disease. Therapeutic stem cell applications may offer hope for these severely ill patients, although some data suggest that cell therapy may have only a limited effect in the elderly, because of the physiological changes that have occurred in the aged myocardium, and by the aged (autologous) stem cells themselves. For elderly patients who remain symptomatic despite intensive medical treatment, autologous bone marrow cells (BMCs) represent a very promising attempt to repopulate lost myocardial tissue (as described below). To intensify the benefit of the autologous stem cell applications in the elderly, the following approaches will all have to be considered in the future: (1) an increased extraction of BM blood and cell number; (2) a pre-treatment of the BM-derived mononuclear cells with specific growth factors in vitro; (3) the injection of a higher amount of regenerative cells; and (4) enhanced ischemia of the myocardium induced by prolonged intracoronary balloon dilation. Therapy with BMCs is ethically justified for the treatment of patients of all ages.

As previously mentioned, cell-based therapy and, in particular, the autologous approach (see below), can represent the unique treatment to apply in old patients or in refractory conditions. A variety of adult stem and progenitor cells from different sources have been examined in recent years for their potential to promote cardiac repair and regeneration. Among these, after the initial characterization of putative

EPCs and their potential to promote cardiac neovascularization and to attenuate ischemic injury, a decade of intense research has examined several novel approaches to promote cardiac repair in adult life. In particular, encouraging data have demonstrated EPC presence in the sites of vascular injury and ischemia. This has led to several preclinical studies being performed in animal models. Promising findings have been obtained. In particular, a favorable improvement in LV function in a rat model of myocardial infarction (MI) after intravenous injection of ex vivo expanded human CD34+ cells has been reported. Furthermore, another study examined the effect of catheter-based intramyocardial transplantation in a swine model of myocardial infarction (MI), providing encouraging outcomes in favoring the application of EPCs as a potential cell therapy in clinical trials. In 2005, Naruse and colleagues carried out a study related to the therapeutic treatment of diabetic neuropathy by in vivo expanded human EPCs, using streptozocin-induced diabetic Nude rats. They developed augmented conduction velocity and ameliorated blood flow of the sciatic nerve. An increased number of microvessels were also observed on the site of EPC injection. These results led to the use of this treatment for cerebrovascular disease. An improvement of neurological functions was reported in chronic cerebral ischemic rats injected with CD34+ HSC cells, including EPCs (Naruse et al. 2005).

The ability of EPCs to expand in cultures under in vitro conditions raises another concern for their therapeutic use. Genetically modified and ex vivo expanded EPCs may become promising new agents. In the Rhesus model, ex vivo CD34+ cell transfection with recombinant non-replicative herpes virus vector and subsequent cell transplantation resulted in the expression of vector genes in angiogenic areas of skin autografts of rhesus macaques. Since CD34+ cells possess a natural angiogenic tropism to injured endothelium, they may serve as ideal candidates for the delivery of genes into areas of angiogenesis (as summarized in Strauer and Steinhoff 2011).

These encouraging data have led to research in clinical trials in order to detect whether EPCs increase endothelial integrity and vascularization at ischemic sites in patients with CVDs. Three different strategies have been principally used. The first strategy consists of the *administration of granulocyte-colony stimulating factor (GCSF)* in the order to determine the recruitment of the patient's own BM resident progenitors. Using this treatment, two preliminary studies demonstrated an increased LV function. This certainly requires confirmation in large studies. The second strategy is *the intracoronary infusion of BM progenitor cells in patients with MI*. This demonstrated positive effects on LV function in three smaller studies. Subsequently, two prospective large trials assessed significant LV function after 4–6 months of administration of BM progenitor cells. Recent and large trials (precisely 10 in number) and large trials confirmed the successfulness and the safety of this procedure with a follow-up over 1.5 years. The third strategy is more invasive and consists of *the direct injection of cells into target tissues*. This treatment (and precisely *transepicardial or transendocardial injection of unfractioned BM cells*) has been performed in patients with diffuse coronary artery disease and intractable angina with no option of recanalization. Ventricular function and physical capacity have been observed to increase, but the small sample size of these studies means that the results need to be confirmed in larger studies. The treatment with *direct administration of EPCs* has been also effectuated in patients with chronic limb ischemia, demonstrating a reduced rate of limb amputation at 3 years of follow-up (as summarized in Strauer and Steinhoff 2011).

However, it is important to be clear that the clinical results obtained in the majority of these studies are limited. EPC infusion did not produce the expected outcome in infarcted patients. Thus, the results obtained to date significantly differ to those from animal studies. For instance, the left ventricular ejection fraction (LVEF) was either unaffected or exhibited only a 5% increase, the minimum change required for long-term survival, upon intracoronary injection of EPCs (sorted as CD133+ cells) into the infarcted myocardium.

As an alternative approach, particularly indicated for old people (as previously mentioned), *autologous cell therapy* has emerged in the last two decades. In order to perform this approach, important prerequisites are needed, including the precise and careful preparation of the cells harvested from the adult BM, the concentration of high cell numbers within the infarction, predominantly in the ischemic border zone, an enhanced migration of stem cells into the apoptotic and necrotic myocardial tissue, and the homing of the injected cells in the damaged myocardium, to avoid the recirculation loss of the injected cells to BM, spleen, liver and lung. To be specific, for cell therapy, 80–250 ml adult BM blood is aspirated from the iliac crest under local anesthesia. In the past, the mononuclear fraction of cells was separated from the whole BM aspirate by density gradient centrifugation using osmolaric media such as ficoll or gelatineoly-succinate (as summarized in Strauer and Steinhoff 2011). However, both methods comprise open preparation procedures and need several washing steps; thus, both techniques need a good manufacturing practice process to produce a quality-controlled cell product and avoid contamination of the end product. That is especially mandatory for further processing of CD133 or CD34 purification of stem cells. Furthermore, both manual preparation protocols take at least 4 h. During cell preparation, viability needs to be determined several times and, finally, must reach approximately 95%. Cell product characterization by fluorescence-activated cell sorting or a cell counter is needed for individual release. Recently, several new automatic systems have been developed to gain nucleated or mononuclear cells from the whole BM aspirate. The advantage of such systems is the possibility to separate the cells in a closed system. In these systems, the cell recovery is higher than with manual preparation, and with the same functional capabilities. Additionally, the preparation time is definitely shorter. The cell preparation and cell application can be completed in one working process, which is considerably cheaper than the conventional BMC preparation procedures. Nowadays, three different separation strategies exist: (1) separation of the total nucleated cells from the BM aspirate; (2) separation of the mononuclear cell fraction; and (3) purification selection of specified stem cells including CD34 or CD133 cells. Most of these automatic separation systems separate different cell populations. Therefore, the clinical specialist has to decide which system fits best for the chosen application and cardiovascular setting. Furthermore, a consensus has to be reached to establish a standard protocol for the characterization and testing of transplantation products in cardiovascular setting and a standard quality of the final cell product.

Concerning cell delivery, different transplantation methods can be used, including intracoronary, endocardial and epicardial intramyocardial transplantation methods in heart disease. After the delivery, the regenerative potential of BM-derived stem cells may be explained by any of the following four mechanisms: (1) direct cell transdifferentiation from BMCs to cardiac myocytes; (2) cytokine-induced myocyte growth and increase of residual viable myocytes (especially in the border zone of the infarcted area); (3) stimulation of intrinsic myocardial stem cells (endogenous stem cells); and (4) induction of cell fusion between transplanted BMCs and resident myocytes, which was taken as an explanation for trans-differentiation. The influence of cytokines has been shown to restore coronary blood vessels and muscle cells after experimental infarction by angiogenesis. BM stem cells express a bounty of cytokines (e.g., vascular endothelial growth factors, insulin-like growth factor, platelet-derived growth factor), thereby stimulating residual normal myocytes for regeneration and proliferation, and intrinsic myocardial stem cells (endogenous stem cells) for cell regeneration and fusion.

Furthermore, ischemic pre-conditioning is important. Stromal derived factor-1 (SDF-1) and its receptor CXCR4 are well established as being essential for the enhancement of hematopoietic progenitor cell recruitment and angiogenesis. The expression of stromal-derived factor-1 is up-regulated during acute ischemia and stimulates the CXCR4 receptor, which is expressed on EPCs and BMCs, thereby acting as a chemotactic and pro-migratory factor. Currently, it is not known exactly how many cells are retained in the myocardium after intracoronary infusion and how many migrate into the border zone. Because MI may be an appropriate stimulus for a stem cell to find its optimum myocardial niche, the ischemia-producing stimulus, for example, by balloon dilation during the BMC infusion (ischemic pre-conditioning), seems to be important for the cells to home into the cardiac endothelium. With respect to obvious differences in the intracoronary delivery techniques used in various publications, the variable outcome of results and therapeutic efficiency may be due to the non-standardized mode of BMC infusion into the coronary circulation. Precise methodological standardization seems to be relevant, both for the effectiveness of stem cell therapy in clinical heart disease and the comparability of multicenter stem cell studies.

The next section reports on the clinical results and indications of autologous therapies.

2.7.2 The Clinical Results and Indications of Autologous Therapies

In acute myocardial infarction, a variety of studies have demonstrated longstanding (up to and over 3 years) improvement of ventricular performance after using stem cell therapy, resulting in an increase in ejection fraction of 3–36% (mean: 11.4%) and a decrease in infarct size of 1–60% (mean: 34%). In most studies, stem cell transplantation was performed in a time frame of 8–14 days after infarction. Although large variability of hemodynamic data after cell therapy exists, there is moderate, but unequivocal improvement of performance of the infarcted heart after stem cell therapy that is quantitatively more than the sum of the interventional measures (e.g., PI and stent) and may be achieved in addition to these therapeutic interventions and to pharmacotherapy. Thus, autologous stem cell therapy represents an innovative and effective procedure for the regeneration of impaired heart muscle in the early phase after the infarct (as summarized in Strauer and Steinhoff 2011).

The reason for the large variety of stem cell effects and for minor or negative results in some studies may be stem cell-related or dependent on different methods for the heart's functional evaluation: for example, by: (1) different methodology of cell preparations associated with altered cell viability; (2) various ages of patients with age-dependent loss of cell viability; (3) non-standardized cell delivery to the heart, especially of the intensity of ischemic pre-conditioning during cell transfer, which represents an important prerequisite for ischemia-induced cell migration; (4) various amount of delivered cells; (5) different times between the acute infarct and stem cell therapy; and (6) non-calculable access to the border zone between the infarct and the unaffected tissue because of vessel occlusion or insufficient intracoronary cell delivery. Moreover, methods for the assessment of ventricular function and perfusion (ventriculography, echocardiography, magnetic resonance imaging, single-positron emission computed tomography) are often not comparably used. This variability of methods is likely to lead to non-uniform and non-standardized cell availability to the damaged area of interest and may impede the comparability of data of various trials. Therefore, exact and comparable methodologies in terms of cell preparation and cell delivery, and of the clinical patient selection procedures are necessary.

In the case of chronic infarction/ischemic heart disease, several clinical studies have revealed beneficial stem cell effects in sub-acute and chronic ischemic heart failure (as summarized in Strauer and Steinhoff 2011). Combined with CABG, the improvement of cardiac function by the use of BMCs has been described as an increase in LVEF of approximately 10%. Studies combining stem cell transplantation with off-pump coronary surgery report similar results, suggesting that cardiac arrest is not mandatory for safe and efficient stem cell implantation. However, these results are always difficult to interpret conclusively without consideration of the revascularization effects. Therefore, recent reports about "stand-alone stem cell treatment" for patients with ischemic heart failure are very interesting. A recent study reported not only a gain in cardiac function but also a clear improvement in quality of life for patients with chronic ischemic heart disease and refractory angina treated after stand-alone BM stem cell injection through lateral mini-thoracotomy (as summarized in Strauer and Steinhoff 2011).

In case of dilated cardiomyopathy, few data have been reported until now on stem cell therapy for dilated cardiomyopathy. The first-in-human study of autologous BM cells in this disease, investigated 44 patients, and the Düsseldorfer ABCD trial investigated 20 patients. In both studies, none of the patients had coronary disease (excluded by angiography) or myocarditis (excluded by endo-myocardial biopsy). In both trials, cell transplantation was performed by the intracoronary administration route in either coronary artery.

In addition, several trials running currently are trying to answer the questions mentioned in the preceding text. Regarding the effect of intracoronary BM progenitor cell infusion in the setting of acute myocardial infarction, placebo-controlled Phase II/III trials such as REGEN-AMI (Bone Marrow Derived Adult Stem Cells for Acute Anterior Myocardial Infarction) are of interest. In the field of surgical cell therapy, the recently launched PERFECT (intramyocardial transPlantation of bonE maRrow stem cells For improvEment of post-inFarct myoCardial regeneraTion, in addition to CABG surgery) study is the first placebo controlled, double-blind, multicenter Phase III trial investigating the effects of intramyocardial BMC injection combined with CABG surgery. Although representing Phase I and II levels, PROMETHEUS (Prospective Randomized Study of Mesenchymal Stem Cell Therapy in Patients Undergoing Cardiac Surgery) is highly interesting because it represents the first-in-human study analyzing the safety and efficacy of intramyocardial injection of mesenchymal stem cells during CABG in patients scheduled for coronary surgery due to ischemic heart disease, as an alternative cell population to the hematopoietic progenitor cell populations mainly used in clinical trials for cardiac regeneration to date. In this respect, the combination treatment of purified EPCs and mesenchymal stem cells has been addressed successfully in a Phase I trial with intramyocardial injection. There are several more interesting trials currently recruiting patients, and results from all of these are needed for a valid evaluation of the gain in cardiac function related to stem cell therapy (as summarized in Strauer and Steinhoff 2011). On the other hand, the growing interest on performing these studies also is linked to the fact that autologous therapy can be effective and safe. No increase of malignant diseases or inadequate progression of coronary artery diseases has been documented. To assess any inflammatory response and myocardial reaction after intracoronary autologous stem cell transplantation, white blood cell count, serum levels of C-reactive protein and of creatine phosphokinase are measured before, during and after treatment, and these data collectively revealed no evidence of inflammation. Neither procedural nor cell-induced complications nor any other type of side effects have occurred so far.

2.8 EPCs as Therapeutic Agents for Cancer: Focus on Anti-VEGF Treatments and Controversies in Their Efficacy

Based on their intrinsic ability to home to tumor sites, EPCs are attractive as cell vectors for targeted cancer gene therapy (Katoh 2013; Marçola and Rodrigues 2015). Scientists have developed genetically engineered EPCs, transfected with vectors encoding some specific antitumor molecules. In that approach, the cells retain their homing properties but lose their capacity to form new blood vessels. In animal models of melanoma, delivery of specific metalloproteinase (MMP-12) by such EPCs has been shown to induce cleavage of molecules that induce tumor progression, thus inhibiting tumor growth, angiogenesis and metastasis. The same type of therapy has also prolonged survival in tumor-bearing mice. The genetically modified EPCs release CD40 ligand that induces the production of tumor necrosis factor and interferon gamma, as well as increasing the activity of caspase-3 and caspase-7, in metastatic lung cancer. In addition, it has been demonstrated that these cells enhance antitumor effects by inhibiting angiogenesis and inducing apoptosis in a murine model of glioma (Katoh 2013; Marçola and Rodrigues 2015).

Other approaches are focused on the use of human/humanized monoclonal antibodies and small-molecule inhibitors preferentially targeting VEGF and VEGFRs. Indeed, they have been developed as anti-angiogenic therapeutics. Bevacizumab (Avastin) is a humanized anti-VEGF monoclonal antibody, which is applied for the treatment of patients with metastatic colorectal cancer, non-small cell lung cancer (NSCLC), ovarian cancer and other types of cancer. Sunitinib (Sutent) and sorafenib (Nexavar) are multi-kinase inhibitors, targeting VEGFR2 and other protein kinases. Sunitinib is applied for the treatment of renal cell carcinoma (RCC) and gastrointestinal stromal tumor (GIST), while Sorafenib is applied for the treatment of RCC and inoperable hepatocellular carcinoma (HCC) (Katoh 2013; Marçola and Rodrigues 2015).

However, this last approach has a very important clinical problem, represented by drug resistance, as a result of clonal expansion or subclonal evolution of tumors as a result of the upregulation of other angiogenic factors, such as Fibroblast growth factor 2 (FGF2). Dovitinib (TKI258), ponatinib (AP24534) and AZD4547 are multi-kinase inhibitors, targeting VEGFR2 and FGFR. FGFRs are aberrantly activated in cancer cells due to gene amplification, translocation, point mutation or other regulatory signaling mechanisms. Since the dual inhibition of VEGFR and FGFR induces anti-tumor effects directly through cancer cells and indirectly through endothelial cells, VEGFR/FGFR dual inhibitors have been investigated in clinical trials for various types of human cancer (Katoh 2013; Marçola and Rodrigues 2015).

As a result, adverse effects associated with the systemic administration of VEGF- or VEGFR-targeting drugs may be observed, including hypertension, proteinuria, cardiac ischemia, cerebral thrombosis, hemorrhage and gastrointestinal perforation (Katoh 2013; Marçola and Rodrigues 2015). Thus, it is important to clarify that anti-VEGF treatments does not produce any objective benefit in a minority of the patients, while, in others, an initial improvement, in the form of tumor shrinkage or stasis is inevitably followed by disease rebound and progression until the patient's death. The failure of anti-VEGF drugs seems to depend on our poor understanding of tumor vascular biology, as recently discussed by Moccia and Poletto (2015) and Moccia et al. (2015). VEGF-dependent angiogenesis does not support neovessel formation in all human cancers, as shown by recent work on GBM and lung adenocarcinoma, which rely heavily on vasculogenic mimicry and vessel cooption, respectively, to gain access to the vascular system (Vasudev and Reynolds 2014). Moreover, it is now evident that tumors exert long-distance systemic effects with the aim of creating an environment that supports survival, proliferation and metastatization of neoplastic cells. Recent evidence indicates that BM-derived cells, such as hematopoietic progenitor cells and endothelial colony forming cells (ECFC), are reprogrammed towards a more aggressive and VEGF-independent phenotype. In particular, Moccia et al. (2014, 2015) clearly showed that ECFCs derived from patients with metastasis are insensitive to VEGF, which might explain why current anti-angiogenic treatments either encounter intrinsic refractoriness from the early beginning or rapidly lead to adaptive resistance. Future studies will have to assess whether anti-VEGF drugs are effective in xenograft murine models of kidney cancer injected with normal. Moreover, it will be important to ascertain whether VEGF fails to activate ECFCs in other types of angiogenic human cancers, such as breast carcinoma. In general, extreme caution is warranted when translating data obtained on healthy BMDCs into clinical application. Based on these findings, it is necessary to examine whether tumor patient-derived hematopoietic cells, as well as the other EPC subtypes described in the literature (e.g., CECs and CACs), become insensitive to VEGF. Gaining access to BM-derived cells isolated from real patients, rather than relying on murine models that cannot recapitulate the complex biology of human tumors, will likely shed novel light on the mechanisms of resistance to anti-angiogenic treatments and disclose alternative, more realistic targets for successful therapies.

Thus, additional studies are needed to investigate efficient EPC treatments in cancer.

2.9 EPCs as Therapeutic Agents for Alzheimer's Disease

As is well recognized, Alzheimer's disease (AD) is characterized by vascular endothelial dysfunction, which has a crucial role in its pathogenesis. Accordingly, subjects affected have decreased circulating levels of EPCs. A beneficial therapeutic approach might consequently be the transplantation of EPCs, as a promising approach for the management of cerebrovascular diseases including ischemic stroke. However, its impact on AD has been poorly described.

Recently, the group of Safar investigated the effects of the transplantation of bone marrow-derived endothelial progenitor cells (BM-EPCs) in repeated scopolamine-induced cognitive impairment, an experimental model that replicates biomarkers of AD. The transplantation was intravenously performed. Transplanted BM-EPCs migrated into the brain of rats and improved learning and memory deficits. In the meantime, they mitigated the deposition of amyloid plaques and associated histopathological alterations. At the molecular level, BM-EPCs blunted the increase of hippocampal amyloid beta protein (Aβ) and amyloid precursor protein (APP), and reinstated the Aβ-degrading neprilysin together with down-regulation of p-tau and its upstream glycogen synthase kinase-3β (GSK-3β). They also corrected the perturbations of neurotransmitter levels including restoration of acetylcholine and associated esterase along with dopamine, GABA, and the neuroexcitatory amino acid, glutamate. Furthermore, BM-EPCs induced behavioral recovery via boosting of vascular endothelial growth factor (VEGF), nerve growth factor (NGF), brain-derived neurotrophic factor (BDNF) and its upstream cAMP response element binding (CREB), suppression of the proinflammatory tumor necrosis factor-α (TNF-α) and interleukin- 1β (IL-1β), and up-regulation of interleukin-10 (IL-10). BM-EPCs also augmented Nrf2 and seladin-1. Generally, these actions were analogous to those exerted by adipose tissue-derived mesenchymal stem cells (AT-MSCs) and the reference anti-Alzheimer's disease drug, donepezil. Thus, for the first time, these researchers highlight the beneficial actions of BM-EPCs against memory deficits and AD-like pathological dysfunction. However, in order to confirm these findings, other studies will be essential (Safar et al., 2016).

2.10 EPCs as Therapeutic Agents for Diabetes

Endothelial dysfunction and endothelial loss are frequently observed in the diabetes setting. In medium-size and larger arteries, this loss of endothelial integrity permits the permeation of atherogenic lipoproteins in close proximity to medial smooth muscle cells. The latter then migrate into the intima, undergoing phenotypic changes that, in conjunction with an associated inflammatory response, lead to the formation of atherosclerotic plaque. In the microvasculature, on the other hand, endothelial loss leads to capillary rarefaction with reduced tissue perfusion and the development of diabetic cardiomyopathy, nephropathy, neuropathy. Although damaged endothelium could potentially be restored, reparative mechanisms also are dysfunctional in diabetes (Yoder 2012; Balistreri et al. 2015).

Cell therapies might certainly be the solution to these changes. Cell treatments might potentiate and induce endothelial renewal and, consequently, may reduce or delay the onset of typical long-term complications related to diabetes. These treatments have been poorly investigated, and could be used in the management of CVDs related to diabetes and, in particular, for diabetic retinopathy (Lois et al. 2014). Drug treatments, including statins and renin-angiotensin system blockers, among others, have been investigated for their role in reducing the typical long-term complications related to diabetes (Gilbert 2013). However, contrasting results have been obtained and further studies are needed.

2.11 Controversies About EPC Use as Optimal Candidates for Cell Therapy

Since their discovery, EPCs have been the object of an intensive investigation and a plethora of clinical applications has been put forward, as reported above. As a result, EPCs have been suggested as potential predictive, diagnostic and prognostic disease biomarkers (see above described), as well as therapeutic agents. These efforts have encouraged researchers in the vision to modulate the onset and progression of age-related diseases and their complications. However, the enthusiasm is actually dampened by a large number of critical viewpoints. In particular, insights into EPC biology are leading several research groups to discuss critical aspects of EPCs and to demonstrate the limitations. Thus, these perspectives reduce the large relevance and potentiality of these cells and co-temporally underline the urgent necessity to move versus standardized and common criteria of research for EPCs. This might reduce the heterogeneity of EPC literature data.

The aspects of EPCs principally discussed by the scientific community are the following:

- *Potential capacity of EPCs in improving* in vivo *or* ex vivo *neovascularization.* In healthy adults, EPCs (as CD34+ CD133+ VEGFR2+ EPCs) represent only 0.0001–0.01% of peripheral blood mononuclear cells (PBMCs) (as described in Chap. 1). These low percentages call into question their impact in pathological or physiological processes. Current evidence reports changes in EPC number and function in several age-related diseases (see above). However, different factors may influence levels and viability of EPCs, including methodological approaches (i.e., the timing and ways of taking samples), detection methods and their protocols, the panel of antibodies used for their phenotypical evaluation, the age of patients and their clinical conditions, and the ethnicity of the populations studied (as described in Chap. 1).
- *EPCs as therapeutic agents.* As reported in detail above, the available clinical studies on EPCs as therapeutic agents of age-related diseases show in complex relative advantageous results. Their validity is limited by different factors: (1) the prevalent evaluation in animal models; (2) the small number of patients

enrolled in the majority of trial human studies, the fact that they are randomized not blinded, and the involvement of only a few centers; (3) the exact phenotypic profile of cells used for the treatments, which is always not indicated or missing; (4) the different administration methods used; and (5) the safety and feasibility of the treatments not proved by long-term follow-up results. Teratoma formation, immunoreactivity, or arrhythmias may represent the adverse effects of these treatments. In addition, there are other limitations in the large-scale clinical use of EPCs. EPCs are relatively rare cells, and expansion of sufficient numbers of subpopulations from peripheral blood is hardly possible. Furthermore, in vitro enumeration of progenitor cells for a sufficient quantity for a therapeutic treatment is associated with changes in phenotype, differentiation and risk of cell senescence, and it may require artificial cell pre-activation or stimulation (as summarized in Balistreri et al. 2015).

- *Lack of standardized criteria and consensus for defining, characterizing and identifying EPCs.* EPCs have been largely described as CD34+ CD133+ VEGFR2+ cells. However, other progenitor populations have been recently considered in EPC studies, including CACs, CECs, CPCs, and CEPs (as described in Chap. 1), playing important roles in tissue neovascularization, but with diverse features. CAC and CEP cells represent variable proportions of CD14+ monocyte cells with different angiogenic properties. Despite their lower in vitro proliferation compared with HSCs or cord stem cells, they seem to have a similar ability to increase neovascularization, as reported in experimental models. This leads to the hypothesis that EPCs might be essentially H-monocyte-derived CD14 + cells with variable expression of CD34, CD133, CD45 and KDR and angiogenesis capacity.

2.12 Conclusions and Perspectives

The biomedical community is pursuing new ways of trying to face the imposing challenge of age-related diseases. In particular, the latest discoveries and advanced knowledge in the fields of stem cell biology (i.e. EPC biology) and their ability to provide a cue for counteracting these diseases are leading numerous researchers to focus their attention on RegMed as a possible new and unique treatment able to repair the organs and system affected by the inevitable ageing process and with limited or assent natural regenerative capacity. Co-temporally, researchers are trying to apply stem cells, such as EPCs, in the clinical context, precisely in the management and outcome of age-related diseases. On the other hand, growing evidence demonstrates alterations in the number and functions of EPCs significantly associated with CVDs, diabetes, AD and cancer, as well as their complications. Such findings support the suggestion of EPCs as potential predictive, diagnostic and prognostic biomarkers and provide optimism regarding a better risk biomarker of the age-related diseases than the classical risk factors. However, many aspects still remain to be clarified. Firstly, it is imperative to verify the validity of the actual data because of the lack of a consensus definition of EPCs themselves and the lack of standardized methodologies for the quantification of these cells. A deep EPC characterization and precise definition is essential and further studies will be necessary to establish standardized criteria for EPC identification, protocol and methods. This may clarify the precise biological role or roles of EPCs and validate their clinical application as both biomarkers to test health of organs and systems, or as candidates for cell therapy.

References

Alba AC, Lalonde SD, Rao V, Walter SD, Guyatt GH, Ross HJ (2013) Changes in circulating progenitor cells are associated with outcome in heart failure patients: a longitudinal study. Can J Cardiol 29:1657–1664

Altabas V (2015) Diabetes, endothelial dysfunction, and vascular repair: what should a diabetologist keep his eye on? Int J Endocrinol 2015:848272. doi:10.1155/2015/848272

Balistreri CR, Buffa S, Pisano C, Lio D, Ruvolo G, Mazzesi G (2015) Are endothelial progenitor cells the real solution for cardiovascular diseases? Focus on controversies and perspectives. BioMed Res Int 835934:1–18

Bogoslovsky (2010) Endothelial progenitor cells correlate with lesion volume and growth in acute stroke. Neurology 75:2059–2062

Braitsch CM, Kanisicak O, van Berlo JH et al (2013) Differential expression of embryonic epicardial progenitor markers and localization of cardiac fibrosis in adult ischemic injury and hypertensive heart disease. J Mol Cell Cardiol 65:108–119

Caballero-Villarraso J et al (2012) Moderate-to-high-intensity training and a hypocaloric Mediterranean diet enhance endothelial progenitor cells and fitness in subjects with the metabolic syndrome. ClinSci (Lond) 123:361–373. doi:10.1042/CS20110477

Caballero S, Hazra S, Bhatwadekar A, Li Calzi S, Paradiso LJ, Miller LP, Chang LJ, Kern TS, Grant MB (2013) Circulating mononuclear progenitor cells: differential roles for subpopulations in repair of retinal vascular injury. Invest Ophthalmol Vis Sci 54(4):3000–3009

Cahill PA, Redmond EM (2016) Vascular endothelium—gatekeeper of vessel health. Atherosclerosis 248:97–109. doi:10.1016/j.atherosclerosis.2016.03.007

Chan YH, Siu CW, Yiu KH, Yiu YF, Lau KK, Lam TH, Lau CP, Tse HF (2013) Prolongation of PR interval is associated with endothelial dysfunction and activation of vascular repair in high-risk cardiovascular patients. J Interv Card Electrophysiol 37(1):55–61

Condorelli RA, Calogero AE, Vicari E, Duca Y, Favilla V, Morgia G, Cimino S, La Vignera S (2013) Endothelial progenitor cells and erectile dysfunction: a brief review on diagnostic significance and summary of our experience. Aging Male 16(2):29–32. doi:10.3109/13685538.2013.789159

Critser PJ, Yoder MC (2010) Endothelial colony-forming cell role in neoangiogenesis and tissue repair. Curr Opin Organ Transplant 15(1):68–72. doi:10.1097/MOT.0b013e32833454b5

Croce G, Passacquale G, Necozione S et al (2006) Nonpharmacological treatment of hypercholesterolemia increases circulating endothelial progenitor cell population in adults. Arterioscler Thromb Vasc Biol 26:e38–e39

Cubbon RM, Kahn MB, Wheatcroft SB (2009) Effects of insulin resistance on endothelial progenitor cells and vascular repair. Clin Sci (Lond) 117:173–190. doi:10.1042/CS20080263

D'Amario D, Leone AM, Iaconelli A et al (2013) Growth properties of cardiac stem cells are a novel biomarker of patients' outcome after coronary bypass surgery. Circulation 129:157–172

de la Puente P, Muz B, Azab F, Azab AK (2013a) Cell trafficking of endothelial progenitor cells in tumor progression. Clin Cancer Res 19:3360–3368. doi:10.1158/1078-0432.CCR-13-0462
de la Puente P, Muz B, Azab F, Azab AK (2013b) Cell trafficking of endothelial progenitor cells in tumor progression. Clin Cancer Res 19(13):3360–3368. doi:10.1158/1078-0432.CCR-13-0462
Devaraj S, Jialal I (2012) Dysfunctional endothelial progenitor cells in metabolic syndrome. Exp Diabetes Res 2012:585018. doi:10.1155/2012/585018
Eirin A, Zhu XY, Woollard JR et al (2013) Increased circulating inflammatory endothelial cells in blacks with essential hypertension. Hypertension 62:585–591
Fadini GP, de Kreutzenberg S, Albiero M, Coracina A, Pagnin E, Baesso I et al (2008) Gender differences in endothelial progenitor cells and cardiovascular risk profile: the role of female estrogens. Arterioscler Thromb Vasc Biol 28:997–1004. doi:10.1161/ATVBAHA.107.159558
Fadini GP, Agostini C, Avogaro A (2010) Autologous stem cell therapy for peripheral arterial disease meta-analysis and systematic review of the literature. Atherosclerosis 209(1):10–17. doi:10.1016/j.atherosclerosis.2009.08.033
Fadini GP, Albiero M, Vigili de Kreutzenberg S, Boscaro E, Cappellari R, Marescotti M et al (2013) Diabetes impairs stem cell and proangiogenic cell mobilization in humans. Diabetes Care 36:943–949. doi:10.2337/dc12-1084
Favero G, Paganelli C, Buffoli B, Rodella LF, Rezzani R (2014) Endothelium and its alterations in cardiovascular diseases: life style intervention. Biomed Res Int. 2014:801896. doi:10.1155/2014/801896
Felice F, Zambito Y, Di Colo G, D'Onofrio C, Fausto C, Balbarini A, Di Stefano R (2012) Red grape skin and seeds polyphenols: evidence of their protective effects on endothelial progenitor cells and improvement of their intestinal absorption. Eur J Pharm Biopharm 80(1):176–184. doi:10.1016/j.ejpb.2011.09.002
Fernández JM, Rosado-Álvarez D, Da Silva Grigoletto ME, Rangel-Zúñiga OA, Landaeta-Díaz LL, Flammer AJ, Gössl M, Widmer RJ, et al (2012) Osteocalcin positive CD133+/CD34−/KDR+ progenitor cells as an independent marker for unstable atherosclerosis. Eur Heart J 33:2963–2969
Flammer AJ, Gössl M, Widmer RJ, Reriani M, Lennon R, Loeffler D, Shonyo S, Simari RD, Lerman LO, Khosla S, Lerman A (2012) Osteocalcin positive CD133+/CD34-/KDR+ progenitor cells as an independent marker for unstable atherosclerosis. Eur Heart J 33(23):2963–2969
Gao DC, Nolan DJ, Mellick AS, Bambino K, McDonnell K, Mittal V (2008) Endothelial progenitor cells control the angiogenic switch in mouse lung metastasis. Science 319 (5860):195–198. doi:10.1126/science.1150224
Gao D, Nolan D, McDonnell K, Vahdat L, Benezra R, Altorki N, Mittal V (2009) Bone marrow-derived endothelial progenitor cells contribute to the angiogenic switch in tumor growth and metastatic progression. Biochim Biophys Acta 1796(1):33–40. doi:10.1016/j.bbcan.2009.05.001
Gilbert RE (2013) Augmenting endothelial repair in diabetes: role of bone marrow-derived cells. Can J Diabetes 37:315–318. doi:10.1016/j.jcjd.2013.06.009
Hayashi S, Sato N, Yamamoto A, Ikegame Y, Nakashima S, Ogihara T, Morishita R (2009) Alzheimer disease-associated peptide, amyloid beta40, inhibits vascular regeneration with induction of endothelial autophagy. Arterioscler Thromb Vasc Biol 29:1909–1915. doi:10.1161/ATVBAHA.109.188516
Hill JM, Zalos G, Halcox JP, Schenke WH, Waclawiw MA, Quyyumi AA, Finkel T (2003) Circulating endothelial progenitor cells, vascular function, and cardiovascular risk. N Engl J Med 348:593–600
Katoh M (2013) Therapeutics targeting angiogenesis: genetics and epigenetics, extracellular miRNAs and signaling networks. Int J Mol Med 32:763–767. doi:10.3892/ijmm.2013.1444
Kong XD, Zhang Y, Liu L, Sun N, Zhang MY, Zhang JN (2011) Endothelial progenitor cells with Alzheimer's disease. Chin Med J (Engl) 124(6):901–906
Lao KH, Zeng L, Xu Q (2015) Endothelial and smooth muscle cell transformation in atherosclerosis. CurrOpinLipidol 26(5):449–456. doi:10.1097/MOL.0000000000000219
Lee ST, Chu K, Jung KH, Park HK, Kim DH, Bahn JJ et al (2009) Reduced circulating angiogenic cells in Alzheimer disease. Neurology 72:1858–1863. doi:10.1212/WNL.0b013e3181a711f4

Lee PS, Poh KK (2014) Endothelial progenitor cells in cardiovascular diseases. World J Stem Cells 6:355–366. doi:10.4252/wjsc.v6.i3.355

Lin CP, Lin FY, Huang PH, Chen YL, Chen WC, Chen HY et al (2013) Endothelial progenitor cell dysfunction in cardiovascular diseases: role of reactive oxygen species and inflammation. Biomed Res Int 2013:845037. doi:10.1155/2013/845037

Ling L, Shen Y, Wang K, et al (2012) Worse clinical outcomes in acute myocardial infarction patients with type 2 diabetes mellitus: relevance to impaired endothelial progenitor cells mobilization. PLoS One 7(11)

Lois N, McCarter RV, O'Neill C, Medina RJ, Stitt AW (2014) Endothelial progenitor cells in diabetic retinopathy. Front Endocrinol (Lausanne) 5:44. doi:10.3389/fendo.2014.00044

Lyden D, Hattori K, Dias S, Costa C, Blaikie P, Butros L, Chadburn A, Heissig B, Marks W, Witte L, Wu Y, Hicklin D, Zhu Z, Hackett NR, Crystal RG, Moore MA, Hajjar KA, Manova K, Benezra R, Rafii S (2001) Impaired recruitment of bone-marrow-derived endothelial and hematopoietic precursor cells blocks tumor angiogenesis and growth. Nat Med 7(11):1194–1201

Madonna R, Novo G, Balistreri CR (2016) Cellular and molecular basis of the imbalance between vascular damage and repair in ageing and age-related diseases: As biomarkers and targets for new treatments. Mech Ageing Dev. 2016 Mar 15. pii:S0047-6374(16)30026-4. doi:10.1016/j.mad.2016.03.005

Marçola M, Rodrigues CE (2015) Endothelial progenitor cells in tumor angiogenesis: another brick in the wall. Stem Cells Int 2015:832649. doi:10.1155/2015/832649

Moccia F, Poletto V (2015) May the remodelling of the Ca(2^+) toolkit in endothelial progenitor cells derived from cancer patients suggest alternative targets for anti-angiogenic treatment? Biochim Biophys Acta 1853(9):1958–73

Moccia F, Zuccolo E, Poletto V, Cinelli M, Bonetti E, Guerra G, Rosti V (2015) Endothelial progenitor cells support tumour growth and metastatisation: implications for the resistance to anti-angiogenic therapy. Tumour Biol 36(9):6603–6614. doi:10.1007/s13277-015-3823-2

Moschetta M, Mishima Y, Sahin I, Manier S, Glavey S, Vacca A et al (2014) Role of endothelial progenitor cells in cancer progression. BiochimBiophys Acta. 1846:26–39. doi:10.1016/j.bbcan.2014.03.005

Naruse K, Hamada Y, Nakashima E, Kato K, Mizubayashi R, Kamiya H et al (2005) Therapeutic neovascularization using cord blood-derived endothelial progenitor cells for diabetic neuropathy. Diabetes 54:1823–1828

Nolan DJ, Ciarrocchi A, Mellick AS, Jaggi JS, Bambino K, Gupta S, et al (2007) Bone marrow-derived endothelial progenitor cells are a major determinant of nascent tumor neovascularization. Genes Develop 21(12):1546–1558.

Oterino A, Toriello M, Palacio E et al (2013) Analysis of endothelial precursor cells in chronic migraine: a case-control study. Cephalalgia 33:236–244

Padfield GJ, Tura-Ceide O, Freyer E, Barclay GR, Turner M, Newby DE, Mills NL (2013) Endothelial progenitor cells, atheroma burden and clinical outcome in patients with coronary artery disease. Heart 99(11):791–798

Peters K, Schmidt H, Unger RE, Kamp G, Pröls F, Berger BJ, Kirkpatrick CJ (2005) Paradoxical effects of hypoxia-mimicking divalent cobalt ions in human endothelial cells in vitro. Mol Cell Biochem 270(1-2):157–166

Regina C, Panatta E, Candi E, Melino G, Amelio I, Balistreri CR, Annicchiarico-Petruzzelli M, Di Daniele N, Ruvolo G (2016) Vascular ageing and endothelial cell senescence: Molecular mechanisms of physiology and diseases. Mech Ageing Dev. 4 May 2016. pii: S0047-6374(16)30059-8. doi:10.1016/j.mad.2016.05.003

Rodiño-Janeiro BK, Paradela-Dobarro B, Castiñeiras-Landeira MI, Raposeiras-Roubín S, González-Juanatey JR, Alvarez E (2013) Current status of NADPH oxidase research in cardiovascular pharmacology. Vasc Health Risk Manag 9:401–428. doi:10.2147/VHRM.S33053

Rodrigo M, Mendelsohn L, Bereal-Williams C et al (2013) Circulating endothelial progenitor cells in adults with sickle cell disease. Pulm Circ 3:448–449

Safar MM, Arab HH, Rizk SM, El-Maraghy SA (2016) Bone marrow-derived endothelial progenitor cells protect against scopolamine-induced alzheimer-like pathological aberrations. Mol Neurobiol 53(3):1403–1418.

Schmidt-Lucke C, Rössig L, Fichtlscherer S, Vasa M, Britten M, Kämper U et al (2005) Reduced number of circulating endothelial progenitor cells predicts future cardiovascular events: proof of concept for the clinical importance of endogenous vascular repair. Circulation 111:2981–2987

Sen S, McDonald SP, Coates PT et al (2011) Endothelial progenitor cells: novel biomarker and promising cell therapy for cardiovascular disease. ClinSci (Lond) 120:263–283

Shimoni S, Bar I, Zilberman L, Goland S, Edri O, Gandelman G, Afek A, Shamiss A, George J (2013) Circulating progenitor and apoptotic progenitor cells in patients with aortic regurgitation. Circ J 77(3):764–771

Strauer BE, Steinhoff G (2011) 10 years of intracoronary and intramyocardial bone marrow stem cell therapy of the heart: from the methodological origin to clinicalpractice. J Am CollCardiol. 58(11):1095–1104. doi:10.1016/j.jacc.2011.06.016

Sugawara J, Mitsui-Saito M, Hoshiai T, Hayashi C, Kimura Y, Okamura K (2005) Circulating endothelial progenitor cells during human pregnancy. J Clin Endocrinol Metab 90:1845–1848

Taguchi A, Matsuyama T, Moriwaki H et al (2004) Circulating CD34-positive cells provide an index of cerebrovascular function. Circulation 109:2972–2975

Tang Z, Wang A, Yuan F, Yan Z, Liu B, Chu JS, Helms JA, Li S (2012) Differentiation of multipotent vascular stem cells contributes to vascular diseases. Nat Commun 3:875. doi:10.1038/ncomms1867

Van Spyk EN, Chun KC, Samadzadeh KM, Peters JH, Lee ES (2013) Increased levels of CD34 + cells are associated in patients with abdominal aortic aneurysms compared with patients with peripheral vascular disease. J Surg Res 184:638–643

Vasa M, Fichtlscherer S, Adler K et al (2001) Increase in circulating endothelial progenitor cells by statin therapy in patients with stable coronary artery disease. Circulation 103:2885–2890

Vasudev NS, Reynolds AR (2014) Anti-angiogenic therapy for cancer: current progress, unresolved questions and future directions. Angiogenesis 17(3):471–494

Vaturi M, Perl L, Leshem-Lev D et al (2011) Circulating endothelial progenitor cells in patients with dysfunctional versus normally functioning congenitally bicuspid aortic valves. Am J Cardiol 108:272–276

Werner N, Nickenig G (2006) Influence of cardiovascular risk factors on endothelial progenitor cells: limitations for therapy? Arterioscler Thromb Vasc Biol 26:257–266

Williamson K, Stringer SE, Alexander MY (2012) Endothelial progenitor cells enter the aging arena. Front Physiol. 20(3):30

Yoder MC (2012) Human endothelial progenitor cells. Cold Spring HarbPerspect Med 2:a006692

Chapter 3
Endothelial Progenitor Cells: A Real Hope or an Unrealizable Dream? Which Measures or Strategies Are Necessary for making EPCs a clinical reality? Focus on a Potential Roadmap

Abstract It is the time to re-examine with new eyes the findings obtained to date on endothelial progenitor cells (EPCs), in order to eliminate the limitations and to shed light on such findings. This revision might be of help for reordering them and consequently to create a clear puzzle. This last might provide evidence relating to important aspects, including defining, isolating and characterizing EPCs by establishing standardized criteria for EPC research; discriminating between appropriate sub-populations for cell therapy; timing; dosing; priming of cells; and the delivery mode for different applications. Furthermore, influencing factors might also be identified, with the aim of focusing resources and efforts; and the problems associated with local retention and fate of cells in the therapeutic target zone might be resolved. Finally, in order to enable the efficacy of cell therapy at all levels, cell priming, bio-nanotechnology and tissue engineering might be used as emerging tools, and, in particular, their use in combination might be very advantageous.

Keywords Re-examination of data on EPCs · To shed light on EPC findings · To create a clear puzzle · New strategies

© Springer International Publishing AG 2017

C.R. Balistreri, *Endothelial Progenitor Cells*, UNIPA Springer Series,
DOI 10.1007/978-3-319-55107-4_3

In the figure (to the right), **Valentin Louis Georges Eugène Marcel Proust** (10 July 1871–18 November 1922), a French novelist, critic and essayist best known for his monumental novel *À la recherche du temps perdu* (*In Search of Lost Time*; earlier rendered as *Remembrance of Things Past*), published in seven parts between 1913 and 1927. He is considered by many to be one of the greatest authors at all time. He also is famous for many quotations, including: "*The real voyage of discovery consists not in seeking new landscapes, but in having new eyes*" (as shown to the left of figure).

This quote can be used as a cue for researchers to shed light on previous findings from cell-based therapies. It is the time for looking such findings with new eyes in order to translate cell-based therapy into real clinical applications. Accordingly, a recent consideration of the scientific community is that cell-based therapy currently is at the phase of translation. However, further time and efforts still are necessary. After the initial hype on cell based therapy, which has led to perform clinical trials, researchers stay in a conclusive stage characterized by more questions than answers. As result, they determine to a step back in the awaited design of therapeutic revolution, which requires an appropriate revision before of a new decisive starting, which leads to real clinical applications.

—*Carmela Rita Balistreri*

3.1 EPCs: A Real Hope or a Unrealizable Dream?

The observations described in Chaps. 1 and 2 provide a critique on the research conducted into EPCs, which seems to underline EPCs as more hype than hope. These observations raise concerns about EPCs as appropriate cell type for angiogenic therapy. In addition, it has been recognized that the levels and functional activity of these cells are lower in patients with cardiovascular diseases (CVDs), diabetes and Alzheimer's disease (AD), further limiting the likelihood of their successful clinical use (see Chap. 2). While true EPCs can be detected in adult blood and isolated from bone marrow (BM), their numbers are very low, making it hard to envisage their successful use for autologous angiogenic therapy (at least until the promise of producing defined EPCs by genetic manipulation of autologous adult somatic cells in order to yield inducible pluripotent cells becomes a practical reality) (see Chap. 2). To further add to these limitations, there are still problems linked to the lack of standardized methodologies and methods for evaluating EPCs in terms of number and functionality. Precisely there are problems in defining, phenotyping and characterizing EPCs and their subpopulations, and consequently discriminating between those sub-populations, which reveal EPCs to be more suitable both as biomarkers of age-related diseases and as optimal candidates for cell therapy. Additional studies are needed, and it certainly is imperative to create a roadmap that establishes agreement among EPC researchers regarding a common and standardized approach to research. As a second measure, a revision of the findings obtained to date may also be of help. In to create a clear puzzle (Fig. 3.1). This last might provide evidence relating to important aspects, including methods of discriminating between appropriate sub-populations for cell therapy and biomarkers of age-related diseases; timing; dosing; priming of cells; and the delivery mode for different applications (Fig. 3.2). Furthermore, influencing factors may also be identified with the aim of evidencing resources and efforts, and the problem associated with local retention and fate of cells in the therapeutic target zone may be resolved. As a final measure of this roadmap, various strategies could be employed for increasing their efficacy their efficacy as therapeutic agents. In this context, cell priming, bio-nanotechnology and tissue engineering may represent emerging tools, and it is likely that they may be merged into a combined biological approach (Fig. 3.3).

The successive paragraph give an overview of these above-mentioned strategies, particularly stressing those used for enhancing EPC efficacy in CVDs, since emerging data on this topic is increasingly being reported in the literature. In particular, some examples regarding optimization of therapeutic EPC efficacy in CVDs using these tools will be described. The fundamental aim of this explanation is to demonstrate that it might be possible to counteract endothelium dysfunction. Realizing this aim may represent the real solution not just for retarding or preventing CVDs, but also all age-related diseases, since the endothelium is the essential component of the stroma of all tissues and organs of human body (as stressed in Chap. 2).

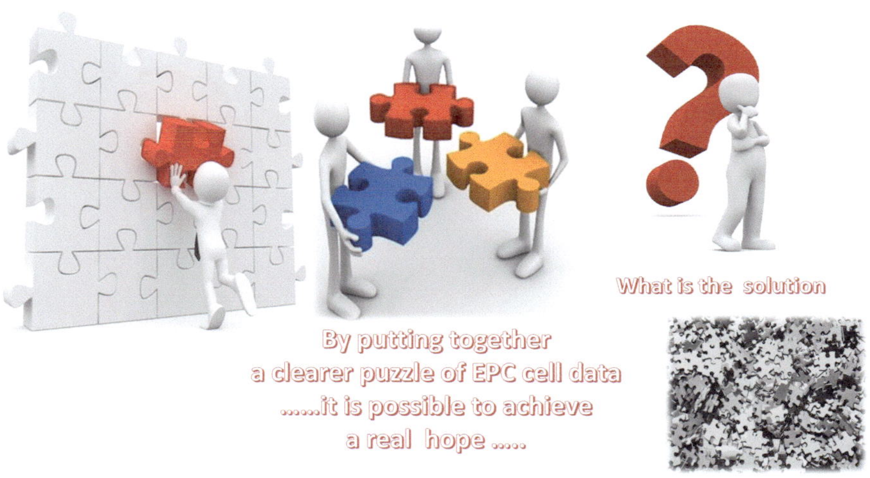

Fig. 3.1 First measure of the roadmap proposed. An accurate revision of the data on EPCs is necessary, because currently there is more confusion than certainty. The solution suggested exists in putting together the valid data obtained to date, in order to create a clear puzzle, This can be of help in providing suggestions and considerations for future investigations

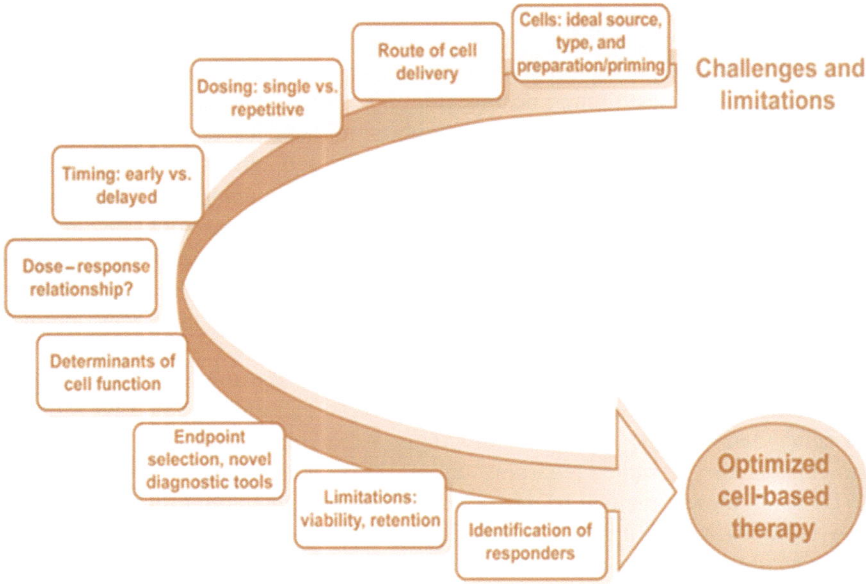

Fig. 3.2 Proposed measures and steps for the optimization of the therapeutic efficacy of EPCs. They provide evidence relating to important aspects, including methods of discriminating between appropriate sub-populations for cell therapy and as biomarkers of age-related diseases; timing; dosing; priming of cells; and the delivery mode for different applications as well as related limitations

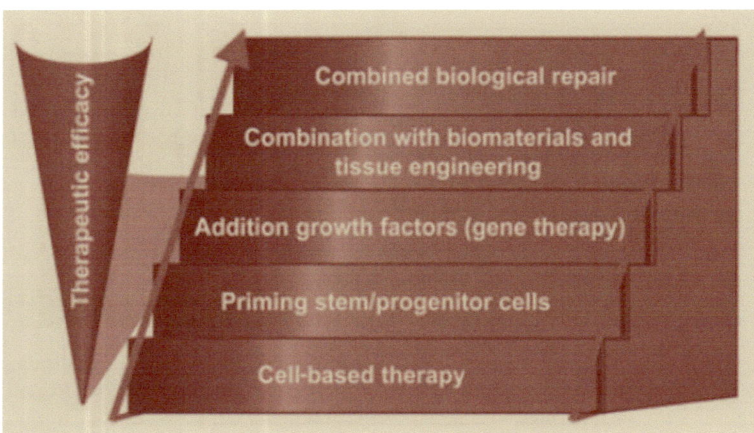

Fig. 3.3 New strategies for improving the therapeutic efficacy of EPCs. In order to optimize the efficacy and potential of EPCs as therapeutic agents and disease biomarkers, new strategies may be used, including cell priming, bionanotechnology, tissue engineering and, eventually, reprogramming. These represent emerging tools, and their combination (as shown in the figure) might produce very advantageous results

3.2 Priming EPCs in Order to Improve Their Therapeutic Efficacy

One of the above-mentioned strategies that has recently emerged, is based on the concept of pre-treating or modifying EPCs before application (*priming*), in order to increase their therapeutic potency. This type of strategy is basically able to modulate cell function and fate from the application of therapeutic cell agent. It specifically influences the following processes: adhesion/transmigration, homing, migration, engraftment, survival, cell–cell interaction, repair capacity, differentiation and retention. To this end, different tools may be used, including drugs, small molecules, naked and vector facilitated plasmids, and epigenetic reprogramming (Penn and Mangi 2008; Chavakis et al. 2010). It has been observed, for example, that the priming of dysfunctional autologous cells from CVD patients via any of these tools may allow for a 'resetting of impaired bio-potency'. Among the multiple targets stemming from pre-clinical evaluation that are now under clinical investigation, some research groups have identified a reduced NO production as an important mechanism limiting the functional repair capacity of EPCs in patients with diabetes or hypertension (Sorrentino et al. 2007; Giannotti et al. 2010). In addition, in the randomized, placebo-controlled Enhanced Angiogenic Cell Therapy in Acute Myocardial Infarction (ENACT-AMI) trial, the therapeutic use of eNOS-overexpressing EPCs is presently being studied in patients with large myocardial infarction (clinical trials.gov: NCT00936819) (Taljaard et al. 2010) As another example of modified stem/progenitor cells, in the Mesenchymal Stem Cells and Myocardial Ischemia II (MESAMI II) trial (clinical trials.gov: registration in process) designed by Roncalli and colleagues, patients with no-option chronic ischemic cardiomyopathy will receive an intra-myocardial injection of Mesenchymal Stem Cells (MSCs) pretreated with melatonin. This concept is based on pre-clinical data showing that pre-treatment with the pineal hormone increases survival, paracrine activity and therapeutic efficiency of MSCs in a rodent model of ischemia (Mias et al. 2008) Another way to boost paracrine and differential cell functions is to pre-stimulate cells before application (Bartunek et al. 2007; Hahn et al. 2008). This concept is currently translated into patients with heart failure from ischemic cardiomyopathy by pre-incubation of BM-derived MSCs with growth factors to guide their transition into cardiopoietic cells (C-Cure trial, clinicaltrials.gov: NCT00810238). Since no clinical study of primed stem cells has been completed yet, it will be interesting to discover whether these strategies will hold promise for clinical applications. Many more targeted cell modifications can be expected to be in the current transition from bench to bedside.

3.3 Bionanotechnology as Support of EPC-Based Therapies

Within damaged tissues or organs, such as infarcted myocardium, there is a very harmful biochemical milieu significantly associated with the loss of tissue architecture and matrix support. These conditions detrimentally alter the micro-environment by modifying the cell–cell and cell–matrix interactions and, in turn, the regulation of dependent pathways that are functionally essential for the homeostasis of transplanted cells. The rapid evolution in the field of bionanotechnology consents to specifically create biomaterials that are able to provide a valid support to transplanted cells within a hostile environment, such as the infarcted myocardium (Laflamme and Murry 2005). In particular, the structure, dimensions and form of constructs are crucial in order to properly mimic the native architecture of extracellular matrix. It is also necessary to define an optimal biomaterial to support cell therapy. A three-dimensional enviroment to increase biomechanical properties of the extracellular matrix is essential and appropriate. This represents a purpose for which controlled organization at the nanoscale is needed (Mooney and Vandenburgh 2008). In some bio-matrices, bioactive signals can be incorporated to specifically modulate EPC biology, also supporting them structurally (Tongers et al. 2009a). The strategy to support cell transfer has rapidly gained attention, triggered by exciting pre-clinical data. In murine models, nanofibers have been used to self-assemble into a matrix, which recruited endogenous progenitors to the myocardium and also supported the successful transplantation of cardiomyocytes, by providing an appropriate microenvironment (Davis et al. 2005). In preliminary studies, biomaterials can be custom-designed to optimally fit the organ-specific microenvironment (Davis et al. 2005) Furthermore, bioactive signals can also be incorporated in some biomaterials to additionally improve cell survival, retention, proliferation and differentiation. In this context, the group of Padin-Iruegas et al. (2009) reported that an insulin-like growth factor carried nanofibers increases EPC-dependent repair of cardiac injury. Furthermore, it has been observed that the combination of a linage-specific optimized and self-assembling nanofiber improves the potency of cell-therapy in case of ischemic tissue repair (Tongers et al. 2009b) In addition, bioactive sequences of biologically attractive paracrine factors (e.g., SDF-1) can more effectively be presented via biomaterials with the aim of recruiting endogenous into or support exogenously applied cells in ischemic myocardium (Segers et al. 2007). Promising results for the role of bionanomaterials in cell-based ischaemic tissue repair are obtaining. However, until now there has not been precedence in humans. Furthermore, for obtaining any clinical translations, several aspects such as dosing, dynamics and kinetics of bioactivity, biodegradability, occurrence and excretion of breakdown products, and immune compatibility have to be carefully addressed.

3.4 Tissue Engineering

Tissue engineering has now been advancing for several years. Regarding replacement of the myocardium, breakthroughs in this field have been hindered by several challenges. While cell seeding of construct backbones has worked well for proliferative and hypoxia-tolerant cells, it has been difficult to achieve tissue-like densities with cells with low proliferation rates such as cardiomyocytes (Laflamme and Murry 2005), although successful generation and transplantation of engineered heart tissue grafts has been reported in small animal models (Zimmermann et al. 2006). Achieving sufficient perfusion of larger constructs is challenging. Passive diffusion is generally tolerated up to a distance of 0.1–0.2 mm. In the case of longer distances, dysfunctionality and necrosis occur as a result of failure of sufficient diffusion (Avgoustiniatos and Colton 1997). At present, bigger, more complex constructs such as the myocardium (1.0–1.5 cm) cannot be maintained by simple diffusion and need anastomosis to a vascular network. Since generation of a stable and functional construct microenvironment remains critical, stem and progenitor cells, such as EPCs, display a strong potential to accelerate tissue engineering for the purpose of seeding, but also for the creation of a vascular networks (Melero-Martin et al. 2008)

3.5 Combined Strategies for Raising Biological Repair

As reported in the majority of studies on EPC therapy, a single-dose transfer of these cells into a deranged environment will likely not be sufficient to 'regenerate' a complex tissue, such as the heart. More recent insights suggest that combined strategies of biological repair more closely resemble native biological mechanisms and, thus, may be more potent than current one-stop, single-cell strategies (Figs. 3.2 and 3.3). Conceptually, the combination of different types of stem/progenitor cells with stromal cells is functionally very attractive aiming at synergisms. Different cell (EPC sub-) populations may exert different regenerative effects. Unselected and selected cells, for example, can show different spatial patterns of homing. Melero-Martin et al. (2008) recently described striking evidence for the synergistic interaction of EPCs with MSCs, which contain relevant pericytes, to create a substantial and functional vascular network in a Matrigel assay that was stable for up to 4 weeks. To add yet another level of complexity, supplemental tools, aimed to improve efficacy of cell transfer, have become available. Engraftment of cardiac-derived stem cells (CSCs) after intra-myocardial application, for example, is improved after sealing the injection site with fibrin glue to attenuate the known leakage of cell suspension (Terrovitis et al. 2009). Low-energy shock wave treatment of the target tissue has been shown to augment recruitment of circulating EPCs by enhancing local expression of chemo-attracting growth factors in mice (Aicher et al. 2006). This technology is supposed to prepare the ischemic target zone for cell transfer and thereby indirectly increase cell homing and retention. In the ongoing Cell-Wave Study (clinical trials.- gov: NCT00326989), extra-corporal shock wave pre-treatment before intracoronary application of BM-derived progenitor cells is already under clinical evaluation in patients with chronic myocardial infarction. In addition, homing of cells via the intracoronary route can be improved by disrupting cell-carrying microbubbles via ultrasound. Without evoking undesired damage, this technology appears to facilitate vascular transmigration by creating capillary pores (Zen et al. 2006; Ghanem et al. 2009). Bearing all the limitations and shortcomings in mind, the idea to realize effective cardiac tissue repair by using only one cell type is likely to be challenging. Depending on further mechanistic and clinical insights, it may predicted that combined, precisely timed, multi-step approaches might be the real solution for having an efficient therapy based on various stem/progenitor and stromal cells, paracrine factors and specifically bioengineered tools for the future of regenerative medicine in order to advance this exciting field to the next level—the level of true tissue and organ repair.

3.6 Reprogramming for Counteracting the Aging and the Senescence of EPCs?

Even though animal aging is an irreversible process at the organism level, as well as at the level of stem and progenitor cells such as EPCs (as reported in Chap. 2), its reversibility is inherent to life, as every generation involves the molecular history of cells from adult individuals being reset to yield new ones free of any age-related burden (Rando and Chang 2012). Thus, fertilization involves the resetting of the aging clock, which can also be achieved experimentally, using approaches aimed at dedifferentiating adult somatic cells to produce embryonic cells, either by somatic cell nuclear transfer or by generation of induced pluripotent stem cells (iPSCs) (Fig. 3.4). These processes illustrate the potential reversibility of aging and fuel the interest to find rejuvenation approaches that do not involve the alterations of the differentiation program intrinsic to the above situations (Rando and Chang 2012). A supply of stem cells with youthful biological properties can have a positive impact on the lifespan and/or health span of the organism. As such, these cells represent preferential targets of potential rejuvenation interventions. It has been reported that cellular senescence, a characteristic feature of aging, is a barrier to cellular reprogramming (Banito et al. 2009). However, the recent implementation of an optimized reprogramming protocol has demonstrated the feasibility of deriving fully functional iPSCs from senescent and centenarian human cells, providing additional evidence of the reversibility of aging at the cellular level (Lapasset et al. 2011).

Independently of cellular dedifferentiation, several approaches have permitted the delay or even reversal of aging in vivo in animal models. This is true in the case of calorie restriction (Cerletti et al. 2012), or the utilization of pharmacological agents such as rapamycin (Selman et al. 2009). NF-kB inhibitors have also been

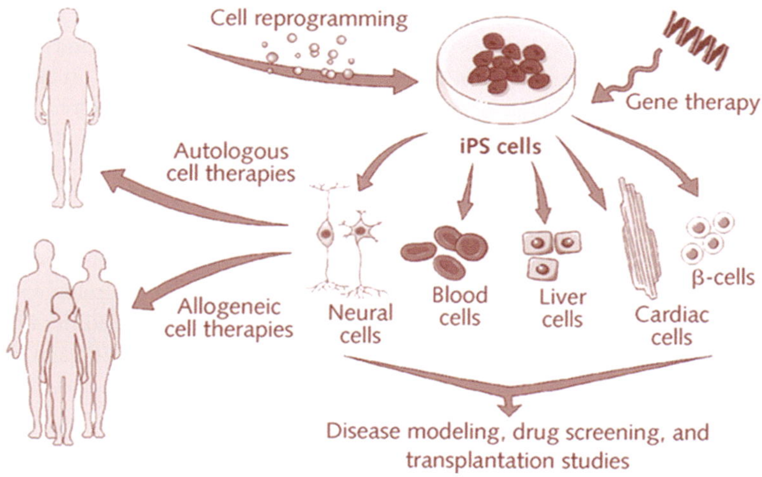

Fig. 3.4 Cell reprogramming: a new source of iPSCs able to give rise to different cell types. Cell reprogramming may allow application of these cells not only in autologous and allogenic cell therapies, but also in studies on disease modeling and drug screening

investigated in this regard on the basis that the activity of this pathway appears to be required to actively maintain the aged state (Adler et al. 2007). Finally, modulation of telomerase activity offers the possibility of acting directly on the biological clock represented by telomere length. Thus, genetic reactivation of telomerase in a telomerase-deficient mouse model reverses the degenerative phenotype in adult mice (Jaskelioff et al. 2011), whereas anti-aging effects have also been obtained with a telomerase-activating drug (de Jesus et al. 2011) and with telomerase gene therapy (de Jesus et al. 2012).

3.7 Conclusions

The observations described in the previous chapters (1 and 2) regarding the critical aspects of EPCs highlight the following considerations: (1) results of earlier studies on EPCs must be re-examined; (2) the impact of EPC subpopulations must be evaluated and considered when future studies are performed; (3) the precise biological role or roles of several EPCs must be clarified prior to their clinical application as both biomarkers or candidates for cell therapy; and (4) EPC definition using surface markers must be re-evaluated taking into account their heterogeneous origin and nature and probably performing a combination of biomolecular assays (see Fig. 3.1).

In order to achieve these aims, it is imperative to create a roadmap that first requires a thorough EPC characterization and precise definition, through performing more and more studies and by establishing standardized criteria for EPC identification, protocol and methods. This might really consent EPC defining and specifying functions. It is likely that a combined and standardized analysis based on cytometric, transcriptomic, proteomic and metabolomic evaluations may be preferentially needed for a definitive and true characterization of these cells, fixing standardized criteria. In addition, in order to optimize the efficacy and potential of EPCs as therapeutic agents and disease biomarkers, new strategies may be utilized, including cell priming, bio-nanotechnology, tissue engineering and, eventually, reprogramming. They represent emerging tools, and their combination might produce very advantageous results (Fig. 3.3). However, the majority of these strategies to date have been conducted in animal models, and not in humans. Thus, additional studies are necessary to inform a real translation to clinical applications of EPCs in humans.

References

Adler AS, Sinha S, Kawahara TL, Zhang JY, Segal E, Chang HY (2007) Motif module map reveals enforcement of aging by continual NF-kappaB activity. Genes Dev 21:3244–3257

Aicher A, Heeschen C, Sasaki K, Urbich C, Zeiher AM, Dimmeler S (2006) Low-energy shock wave for enhancing recruitment of endothelial progenitor cells: a new modality to increase efficacy of cell therapy in chronic hind limb ischemia. Circulation 114:2823–2830

Avgoustiniatos ES, Colton CK (1997) Effect of external oxygen mass transfer resistances on viability of immunoisolated tissue. Ann NY Acad Sci 831:145–167

Banito A, Rashid ST, Acosta JC, Li S, Pereira CF, Geti I, Pinho S, Silva JC, Azuara V, Walsh M, Vallier L, Gil J (2009) Senescence impairs successful reprogramming to pluripotent stem cells. Genes Dev 23(18):2134–2139. doi:10.1101/gad.1811609

Bartunek J, Croissant JD, Wijns W, Gofflot S, de Lavareille A, Vanderheyden M, Kaluzhny Y, Mazouz N, Willemsen P, Penicka M, Mathieu M, Homsy C, De Bruyne B, McEntee K, Lee IW, Heyndrickx GR (2007) Pretreatment of adult bone marrow mesenchymal stem cells with cardiomyogenic growth factors and repair of the chronically infarcted myocardium. Am J Physiol Heart Circ Physiol 292:H1095–H1104

Cerletti M, Jang YC, Finley LW, Haigis MC, Wagers AJ (2012) Short-term calorie restriction enhances skeletal muscle stem cell function. Cell Stem Cell 10:515–519

Chavakis E, Koyanagi M, Dimmeler S (2010) Enhancing the outcome of cell therapy for cardiac repair: progress from bench to bedside and back. Circulation 121:325–335

Davis ME, Motion JP, Narmoneva DA, Takahashi T, Hakuno D, Kamm RD, Zhang S, Lee RT (2005) Injectable self-assembling peptide nanofibers create intramyocardial microenvironments for endothelial cells. Circulation 111:442–450

de Jesus BB, Schneeberger K, Vera E, Tejera A, Harley CB, Blasco MA (2011) The telomerase activator TA-65 elongates short telomeres and increases health span of adult/old mice without increasing cancer incidence. Aging Cell 10:604–621

de Jesus BB, Vera E, Schneeberger K, Tejera AM, Ayuso E, Bosch F, Blasco MA (2012) Telomerase gene therapy in adult and old mice delays aging and increases longevity without increasing cancer. EMBO Mol Med 4(8):691–704. doi:10.1002/emmm.201200245

Ghanem A, Steingen C, Brenig F, Funcke F, Bai ZY, Hall C, Chin CT, Nickenig G, Bloch W, Tiemann K (2009) Focused ultrasound-induced stimulation of microbubbles augments site-targeted engraftment of mesenchymal stem cells after acute myocardial infarction. J Mol Cell Cardiol 47:411–418

Giannotti G, Doerries C, Mocharla PS, Mueller MF, Bahlmann FH, Horvath T, Jiang H, Sorrentino SA, Steenken N, Manes C, Marzilli M, Rudolph KL, Luscher TF, Drexler H, Landmesser U (2010) Impaired endothelial repair capacity of early endothelial progenitor cells in prehypertension: relation to endothelial dysfunction. Hypertension 55:1389–1397

Hahn JY, Cho HJ, Kang HJ, Kim TS, Kim MH, Chung JH, Bae JW, Oh BH, Park YB, Kim HS (2008) Pre-treatment of mesenchymal stem cells with a combination of growth factors enhances gap junction formation, cytoprotective effect on cardiomyocytes, and therapeutic efficacy for myocardial infarction. J Am Coll Cardiol 51:933–943

Jaskelioff M, Muller FL, Paik JH, Thomas E, Jiang S, Adams AC, Sahin E, Kost-Alimova M, Protopopov A, Cadiñanos J, Horner JW, Maratos-Flier E, Depinho RA (2011) Telomerase reactivation reverses tissue degeneration in aged telomerase-deficient mice. Nature 469 (7328):102–106. doi:10.1038/nature09603

Laflamme MA, Murry CE (2005) Regenerating the heart. Nat Biotechnol 23:845–856

Lapasset L, Milhavet O, Prieur A, Besnard E, Babled A, Aït-Hamou N, Leschik J, Pellestor F, Ramirez JM, De Vos J, Lehmann S, Lemaitre JM (2011) Rejuvenating senescent and centenarian human cells by reprogramming through the pluripotent state. Genes Dev 25 (21):2248–2253. doi:10.1101/gad.173922.111

Melero-Martin JM, De Obaldia ME, Kang SY, Khan ZA, Yuan L, Oettgen P, Bischoff J (2008) Engineering robust and functional vascular networks in vivo with human adult and cord blood-derived progenitor cells. Circ Res 103:194–202

Mias C, Trouche E, Seguelas MH, Calcagno F, Dignat-George F, Sabatier F, Piercecchi-Marti MD, Daniel L, Bianchi P, Calise D, Bourin P, Parini A, Cussac D (2008) Ex vivo pretreatment with melatonin improves survival, proangiogenic/mitogenic activity, and efficiency of mesenchymal stem cells injected into ischemic kidney. Stem Cells 26:1749–1757

Mooney DJ, Vandenburgh H (2008) Cell delivery mechanisms for tissue repair. Cell Stem Cell 2:205–213

Padin-Iruegas ME, Misao Y, Davis ME, Segers VF, Esposito G, Tokunou T, Urbanek K, Hosoda T, Rota M, Anversa P, Leri A, Lee RT, Kajstura J (2009) Cardiac progenitor cells and biotinylated insulin-like growth factor-1 nanofibers improve endogenous and exogenous myocardial regeneration after infarction. Circulation 120:876–887

Penn MS, Mangi AA (2008) Genetic enhancement of stem cell engraftment, survival, and efficacy. Circ Res 102:1471–1482

Rando TA, Chang HY (2012) Aging, rejuvenation, and epigenetic reprogramming: resetting the aging clock. Cell 148:46–57

Segers VF, Tokunou T, Higgins LJ, MacGillivray C, Gannon J, Lee RT (2007) Local delivery of protease-resistant stromal cell derived factor-1 for stem cell recruitment after myocardial infarction. Circulation 116:1683–1692

Selman C, Tullet JM, Wieser D, Irvine E, Lingard SJ, Choudhury AI, Claret M, Al-Qassab H, Carmignac D, Ramadani F, Woods A, Robinson IC, Schuster E, Batterham RL, Kozma SC, Thomas G, Carling D, Okkenhaug K, Thornton JM, Partridge L, Gems D, Withers DJ (2009) Ribosomal protein S6 kinase 1 signaling regulates mammalian life span. Science 326 (5949):140–144. doi:10.1126/science.1177221

Sorrentino SA, Bahlmann FH, Besler C, Muller M, Schulz S, Kirchhoff N, Doerries C, Horvath T, Limbourg A, Limbourg F, Fliser D, Haller H, Drexler H, Landmesser U (2007) Oxidant stress impairs in vivo re-endothelialisation capacity of endothelial progenitor cells from patients with type 2 diabetes mellitus: restoration by the peroxisome proliferator-activated receptor-gamma agonist rosiglitazone. Circulation 116:163–173

Taljaard M, Ward MR, Kutryk MJ, Courtman DW, Camack NJ, Goodman SG, Parker TG, Dick AJ, Galipeau J, Stewart DJ (2010) Rationale and design of enhanced angiogenic cell therapy in acute myocardial infarction (ENACT-AMI): the first randomized placebo-controlled trial of enhanced progenitor cell therapy for acute myocardial infarction. Am Heart J 159:354–360

Terrovitis J, Lautamaki R, Bonios M, Fox J, Engles JM, Yu J, Leppo MK, Pomper MG, Wahl RL, Seidel J, Tsui BM, Bengel FM, Abraham MR, Marban E (2009) Noninvasive quantification and optimization of acute cell retention by in vivo positron emission tomography after intramyocardial cardiac-derived stem cell delivery. J Am Coll Cardiol 54:1619–1626

Tongers J, Webber MJ, Losordo DW (2009a) Bioengineering to enhance progenitor cell therapeutics. Tex Heart Inst J 36:140–144

Tongers J, Webber MG, Renault M-A, Roncalli JG, Jujo K, Wu X, Klyachko E, Thorne T, Stupp SI, Losordo DW (2009b) Abstract 4931: RGDS-epitope presenting peptide amphiphile nanofibers enhance therapeutic potency of cell-based strategies in ischemic tissue. Circulation 120:S1023-c-1024

Zen K, Okigaki M, Hosokawa Y, Adachi Y, Nozawa Y, Takamiya M, Tatsumi T, Urao N, Tateishi K, Takahashi T, Matsubara H (2006) Myocardium-targeted delivery of endothelial progenitor cells by ultrasound-mediated microbubble destruction improves cardiac function via an angiogenic response. J Mol Cell Cardiol 40:799–809

Zimmermann WH, Melnychenko I, Wasmeier G, Didié M, Naito H, Nixdorff U, Hess A, Budinsky L, Brune K, Michaelis B, Dhein S, Schwoerer A, Ehmke H, Eschenhagen T (2006) Engineered heart tissue grafts improve systolic and diastolic function in infarcted rat hearts. Nat Med 12:452–458

Appendix

Aubrey de Grey

In the figure (to the left) **Aubrey de Grey** (born 20 April 1963), *an English author and biomedical gerontologist. He is currently the Chief Science Officer of the SENS Research Foundation. He is editor-in-chief of the academic journal Rejuvenation Research, author of The Mitochondrial Free Radical Theory of Aging (1999) and co-author of Ending Aging (2007). He is known for his view that medical technology may enable human beings alive today to live indefinitely. De Grey's research focuses on whether regenerative medicine can prevent the aging process. He works on the development of what he calls "Strategies for Engineered Negligible Senescence" (SENS), a collection of proposed techniques to rejuvenate the human body and to stop aging. To this end, he has identified seven types of molecular and cellular damage caused by essential metabolic processes. SENS is a proposed panel of therapies designed to repair this damage.*

Aubrey has become the leading prophet of the radical life extension movement. To adherents, he has become a messiah figure as well; the long Jesus-like beard serves him well.

No other speaker trumps his passion for the subject of aging. Instead of searching for pragmatic methods to slow aging using current technology, de Grey has his sights set on the future.

His primary message is that, "*even with all the major advances in regenerative medicine and biotech, we are still far from the goal of repairing damage accumulated from aging*". He says, "*I accept the difficult of what you cannot yet change. But do not accept the impossibility of ever changing it*" (as shown to the right of

© Springer International Publishing AG 2017

C.R. Balistreri, *Endothelial Progenitor Cells*, UNIPA Springer Series,
DOI 10.1007/978-3-319-55107-4

figure). Funding is insufficient to benefit people living today; it needs to be ramped up dramatically. Business will only act if there is sufficient demand, along with promise of profitability. Aubrey's goal is to convince scientists, decision makers, and the public at large that the answer to aging will materialize if we want it badly enough. The time has come to vote with our voices and our wallets.

Similarly to Aubrey's message, the author's final message of this monograph is the following:

> "*We must not surrender in the face of the many difficulties that we face in performing our research ... but above all we must not forget that the ultimate goal of our experiments should be for the good of man and not the love of success ... motivated exclusively by this... perseverance, collaboration and honest activity, together with the necessary funding, will allow us to convert assumptions into certainties*"
>
> —Carmela Rita Balistreri

MIX
Papier aus verantwortungsvollen Quellen
Paper from responsible sources
FSC® C105338
www.fsc.org

If you have any concerns about our products,
you can contact us on
ProductSafety@springernature.com

In case Publisher is established outside the EU,
the EU authorized representative is:
Springer Nature Customer Service Center GmbH
Europaplatz 3, 69115 Heidelberg, Germany

Printed by Libri Plureos GmbH
in Hamburg, Germany